STATISTICS

from

SCRATCH

I see that groping is down this month. Had a few days off then, did we?

STATISTICS

from SCRATCH

For Health Care Professionals

David Bowers

The Nuffield Institute for Health,
University of Leeds, and Department
of Liaison Psychiatry, Leeds General Infirmary, UK

JOHN WILEY & SONS
Chichester · New York · Weinheim · Brisbane · Singapore · Toronto

National 01243 779777
International (+44) 1243 779777
e-mail (for orders and customer service enquiries): cs-books@wiley.co.uk
Visit our Home Page on http://www.wiley.co.uk
or http://www.wiley.com

Reprinted November 1997, September 1998

Other Wiley Editorial Offices

John Wiley & Sons, Inc., 605 Third Avenue,
New York, NY 10158-0012, USA

VCH Verlagsgesellschaft mbH, Pappelallee 3
D-69469 Weinheim, Germany

Jacaranda Wiley Ltd, 33 Park Road, Milton,
Queensland 4064, Australia

John Wiley & Sons (Asia) Pte Ltd, 2 Clementi Loop #02-01,
Jin Xing Distripark, Singapore 129809

John Wiley & Sons (Canada) Ltd, 22 Worcester Road,
Rexdale, Ontario M9W 1L1, Canada

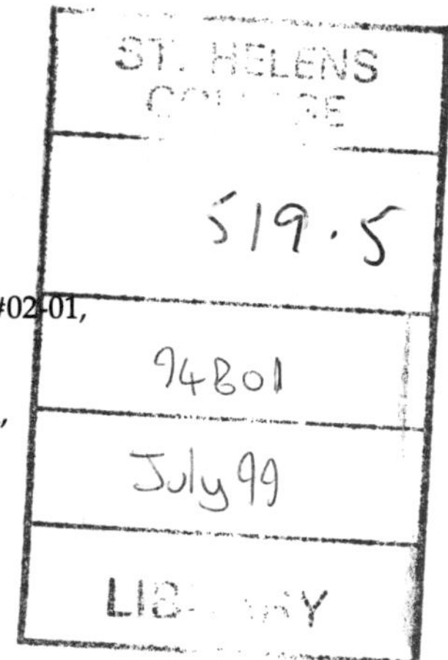

Library of Congress Cataloging-in-Publication Data

Bowers, David.
Statistics further from scratch / David Bowers.
p. cm.
Includes bibliographical references (p.) and index.
ISBN 0-471-97136-7 (pbk. : alk. paper)
1. Medical statistics. I. Title.
RA409.B674 1997
519.5'024'61—dc21 96–47783
CIP

British Library Cataloguing in Publication Data

A catalogue record for this book is available from the British Library

ISBN 0-471-97136-7

Typeset in 10/11½pt Palatino from the author's disks by Dorwyn Ltd, Rowlands Castle, Hants
Printed and bound in Great Britain by Bookcraft Ltd, Midsomer Norton, Somerset
This book is printed on acid-free paper responsibly manufactured from sustainable forestation, for which at least two trees are planted for each one used for paper production.

CONTENTS

ACKNOWLEDGEMENTS

I am very grateful to all the copyright holders who have given me permission to reproduce previously published material:

The *British Medical Journal* for Tables 2.1, 2.2, 2.6, 2.8, 3.5, 3.6, 4.6, 5.9, 5.10, 5.12, and 5.13, copyright 1993, 1994, 1995 the British Medical Journal. *Physiotherapy* for Tables 2.5, 5.7, and 6.10, copyright 1992, 1993. *The Lancet* for Tables 3.7, 4.2, 5.1, and 8.2, copyright 1994, 1995 The Lancet Ltd. The *British Journal of Psychiatry* for Table 4.5, copyright 1995. The *British Journal of Clinical Psychology* for Table 5.14, copyright 1995. The *Journal of Human Nutrition & Dietetics* for Tables 6.7, 6.9, and 8.5, copyright 1994. *Quality in Health Care* for Table 8.4, copyright 1995. Full source details for each item are given in the references.

1

BACK TO BASICS

A BRIDGE OVER TROUBLED WATERS

This book is a follow-on to a companion volume, *Statistics from Scratch* (John Wiley, 1996). In that first book I dealt with the basic ideas underlying what is generally known as *descriptive statistics*, so this book assumes that you are reasonably familiar with these ideas. However, I thought it might be useful to provide a short bridging chapter to act as a reminder of the more important of these concepts. We use descriptive statistics to describe the principal features of a set of *sample* data. Usually this means determining the numeric value of two important summary measures: (i) a measure of *location* or central tendency (either the mean, the median or the mode); and (ii) a measure of *spread* or dispersion (either the range, the interquartile range, or the standard deviation). Collectively, these numeric values are known as the *sample statistics*. In addition, we will also want to know something about the *shape* of the distribution of sample values. Is the distribution symmetric or are the values skewed (and if skewed, in what direction)?

Importantly, we do not attempt to generalise from what we discover about the sample to the broader population from which the sample was taken. Descriptive statistics confines itself to the sample data.

As I suggested in the earlier book, in practice we are likely to want to use what we discover about the sample as a basis for discovering something about the main features, or *parameters*, of the *sampled population*. In other words, we use the values of the sample statistics (for example, the sample mean) to make an informed guess as to the value of the corresponding population *parameter* (for example the population mean). This process, i.e. generalising from the sample to the population, is known as *inferential statistics* (or statistical inference) and is illustrated schematically in Figure 1.1. Statistical inference forms the subject matter of this book.

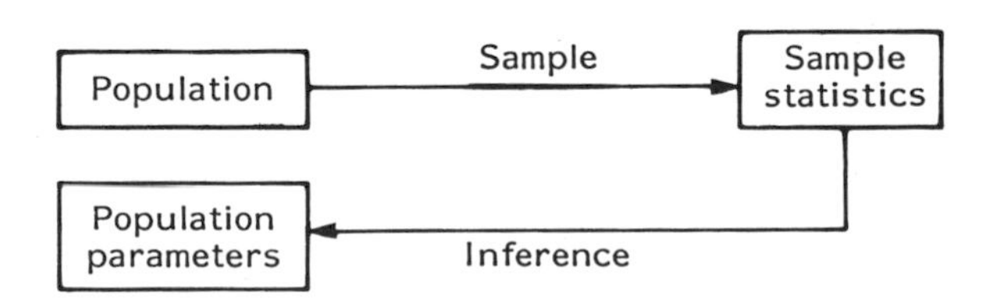

Figure 1.1: The process of inferential statistics

TYPES OF VARIABLES

One of the most fundamental concepts in both descriptive and inferential statistics relates to types of variables. An understanding of the differences between nominal, ordinal and numeric (also known as interval/ratio) variables is crucial, not only in descriptive statistics, but for most of what is to follow here. Variables come in three forms: *nominal*, *ordinal* and *metric*.

Nominal Variables

Nominal variables are used simply to classify or *categorise* data (for this reason nominal variables are often referred to as categorical variables). Examples are blood types (with categories labelled A or B or O or AB, etc.), sex (male or female), ethnicity (Afro-Caribbean, European, Asian, etc.), and so on. The ordering of the categories is completely *arbitrary*; for example, it doesn't matter whether we order blood type categories as A, AB, O, B or B, A, AB, O, or in any other way. Category labels are invariably *alphabetic*, as in the above examples, and have no numeric content. It follows that we can't apply the rules of arithmetic to them. We can't, for example, calculate the *average* blood type of a group of patients, or their average sex. Note that ICD-10 Codes and the Sections of the Mental Health Act, and such like variables, although seemingly "numeric", could easily be replaced with alphabetic characters, and are clearly nominal in nature.

Ordinal Variables

Ordinal variables, too, are used to categorise or classify data. However, unlike nominal variables, there is an inherent or natural *ordering* to the categories. For example, in a patient satisfaction questionnaire the responses can be "very unsatisfied", "unsatisfied", "satisfied", "very satisfied". Or in the progress of breast cancer, Stage I, Stage II, Stage III, Stage IV, etc. With an ordinal variable, the category labels may be either alphabetic, as in the above examples, or "numeric". Examples are Apgar scores of 0, 1, 2, 3, 4, up to 10; or the Glasgow Coma scores, 3, 4, 5, up to 15; or the Barthel Activities of Daily Living scores of between 0 and 20. Most "numeric" ordinal data arise from measuring scales such as these, but can also arise from visual analogue scales (as used in pain measurement, for example).

The reason why I have enclosed the word numeric in quotes above is that these numeric values are not true numbers. We can't say that a baby with an Apgar score of 8 is exactly *twice* as healthy as one with a score of 4, or that the difference in health between a baby with a score of 5 and one with a score of 6 is the same as the difference between a baby with a score of 6 and one with a score of 7. All we can say is that the higher the score, the healthier the baby. (No more than we can say that the difference in level of satisfaction between a patient returning a score "satisfied" and one returning a score of "very satisfied" is the same as that between scores of "very unsatisfied" and "unsatisfied".) This means that, as with nominal data, we cannot apply the rules of arithmetic to "numeric" ordinal scores. One consequence is that we should not calculate the *mean* of a set of numeric ordinal scores (because this would mean first *adding* the values and then *dividing* by their number, two arithmetic operations not appropriate with data such as this). The median (or the mode) would be more appropriate as a measure of average. This is an important point. Data which arise from the measurement of both nominal and ordinal variables are known as *qualitative* data.

Metric Variables

Metric (also known as interval/ratio) variables are used to categorise and order data. But in addition the distance or difference between each category value is the *same*. For example, a patient who waits 20 minutes before being seen in an outpatient clinic does wait *exactly* twice as long as a patient who waits 10 minutes, and the length of time between a patient who waits 10 minutes and one who waits 11 minutes is exactly the same as the difference between patients waiting 20 minutes and 21 minutes. So metric data are truly numeric and all the rules of arithmetic (including addition and division) can be applied to such data, and it would be appropriate to calculate the mean waiting time for a group of patients.

Metric data always have *units* attached. Examples are units of time (seconds, minutes, days, etc.), units of weight (grams and kilograms), units of distance or width (centimetres and metres), units of volume (cc's and litres), units of density (grams per litre, etc.), and so on, or numbers or counts of things (numbers of patients, of pressure sores, of deaths, etc.). Metric data are also known as *quantitative* data.

DISCRETE VERSUS CONTINUOUS DATA

Variables and data can also be classified as being *discrete* or *continuous*. Nominal and ordinal data (from qualitative variables) are invariably discrete, metric data may be discrete or continuous. A discrete variable is one which takes a *countable, limited* or *finite* number of values. For example, the number of pressure sores a patient has might take the values 0, 1, 2, 3, 4, and so on; but we know somewhere there is an upper limit to the number. Consider the number of patients seen in an A&E department in any year. The number of values here may range up to tens of thousands (or more), but the number, even though large, is limited. On the other hand, a continuous variable can take an unlimited or infinite number of values, which cannot be counted. For example, the weight of a patient might be 70.0 kg or 70.1 kg or 70.05 kg or 70.056 kg or 70.0561 kg, and so on *ad infinitum*, the value limited only by the accuracy of the instrument used to measure weight. As a more mundane example, if I showed you a six-pack (closed) egg box and asked you to tell me the possible *number* of eggs in the box you would know there were only seven possible values (0, 1, 2, 3, 4, 5 or 6). But if I asked you to write down the *weight* of the egg box, it couldn't be done since the number of possible values is uncountable (given enough decimal points on the weighing scales). As in this example, discrete variables usually *count* things, whereas continuous variables usually *measure* things.

To reiterate what I said above, it is important to be able to identify correctly what type of variable you are working with since the choice of a suitable descriptive measure (of location or spread) depends crucially on getting this right. As a rule of thumb, the mean and standard deviation are used with quantitative (i.e. metric) variables, unless the distribution is markedly skewed (in which case the median and interquartile range should be used). With qualitative variables, the mode or median and the interquartile range are the most appropriate.

Identification of variable type (and the shape of the distribution of values) is equally important in statistical inference. As we shall see, the most appropriate hypothesis test or confidence interval can only be decided once we know what sort of variable we are dealing with.

THE NORMAL DISTRIBUTION

I referred above to the importance of the shape of the distribution of the values of the variable in the population. If the sample is large enough we can get some idea of the shape by plotting the sample values. If the distribution has mostly lowish values with a few high values this is described as being right or positively skewed. If it has mostly largish values with a few small values, this is known as a left or negatively skewed distribution. If the values are evenly spread throughout the range of possible values the distribution is symmetric.

One symmetric distribution of particular interest is the *Normal* distribution. When plotted this appears as a smooth, bell-shaped curve, whose actual shape is determined by the values of the mean and standard deviation of the data. Any Normal distribution has a number of interesting properties, chief among which for our purposes are the *area* properties. In simple terms this means that fixed areas under the curve (and hence a fixed proportion of the data) lie within a given number of standard deviations of the mean of the distribution*.

For example, about two-thirds of the observations will lie within plus or minus one standard deviation from the mean, about 95% will lie within two standard deviations of the mean, and about 99% within three standard deviations of the mean. Thus, if we knew that the age of men in a population was Normally distributed with a mean of 40 years and a standard deviation of 10 years, we could say that about two-thirds of the men would be aged between 30 (40 – 10) and 50 (40 + 10). To determine these proportions exactly for any distribution with known mean and standard deviation we can make use of a table showing the values between the mean and any given value of the variable in question. This table (Table A1 in the Appendix) contains the values (commonly known as z values) of what is known as the *standardised Normal distribution*, together with the corresponding areas under the standard Normal curve.

PARAMETRIC VERSUS NON-PARAMETRIC STATISTICS

A second major consideration in deciding on the most appropriate approach in statistical inference (which I did not allude to in *Statistics from Scratch*) is the choice between *parametric* and *non-parametric* methods. The choice between

* See Chapter 5 in *Statistics from Scratch* for a detailed discussion of the Normal distribution and the use of *standard Normal distribution*.

these two alternatives depends on the shape of the distribution of the variable in question in the population. If the population distribution is *Normal*, then a parametric method is appropriate (for example, methods based on the t distribution, discussed in the next chapter). If the population distribution is *not* Normal, for example if the distribution is significantly skewed, then a non-parametric method should be used (non-parametric methods are first discussed in Chapter 3).

In practice, since we cannot observe the population, a decision about the shape of the population distribution must be made on the basis of the shape of the distribution of the *sample* data. This means that the sample data should always be plotted so that its shape can be examined. This presupposes that the sample is big enough for a reasonably reliable view to be taken. If the sample is too small this will not be possible (or may lead to a misleading conclusion). In these circumstances you should play safe and use a non-parametric method (unless there is strong evidence of Normality from other sources, perhaps a previous larger sample from the same population).

Broadly speaking, the process of statistical inference, that is discovering something about the characteristics of the population (i.e. the value of the population parameters) from the sample statistics, takes two separate but related forms, estimation and hypothesis testing.

- *Estimation*. If we have no prior idea what the value of the population parameter in question is we use the estimation approach. For example, we can use the sample mean to make an informed guess as to the value of the population mean.
- *Hypothesis testing*. If we do have some prior idea about the possible value of a population parameter, we use the hypothesis testing approach. Thus we might see if the value we get for a sample mean supports (or does not support) our prior idea as to the value of the population mean.

In the health and human sciences there has been in recent years a shift away from the once dominant position of hypothesis testing in statistical inference towards the use of estimation. The reasons for this will become clear during the discussion of both approaches in the rest of this book, but essentially estimation is more *informative* than hypothesis testing. For this reason you should, whenever practicable, use estimation rather than hypothesis testing to analyse your results (or use both). Most major computer statistics packages (Minitab[1], SPSS[2], EPI[3], CIA[4], etc.) offer a choice of either approach for the more commonly used procedures. In some cases, however, particularly with non-parametric methods, the estimation approach may not be readily available. The chi-squared test is a good example of this; the equivalent estimation of differences in proportions is still not often used.

In the next chapter I will start with a discussion of the estimation approach applied to the estimation of the population mean.

REFERENCES

1. Minitab® for Windows™. Minitab Inc., 3081 Enterprise Drive, State College, PA 16801-3008, USA.
2. SPSS® for Windows™. SPSS Inc., 444 N. Michigan Avenue, Chicago, IL 60611, USA.

3. EPI Info. Centre for Disease Control (CDC), Atlanta, GE 30333, USA. Supplier in the UK is the Public Health Laboratory Service, 61 Collindale Avenue, London NW9 5EQ. Tel: 0181 200 6868.
4. Confidence Interval Analysis (CIA), version 0.5 (1989). Gardner, M., Gardner, S. B. and Winter, P. D., British Medical Journal, Tavistock Square, London WC1H 9JR.

2

ESTIMATING THE POPULATION MEAN

❑ FROM SAMPLES TO POPULATIONS ❑ POPULATION PARAMETERS AND SAMPLE STATISTICS ❑ PROBABILITY ❑ THE IDEA OF ESTIMATION ❑ POINT VERSUS INTERVAL ESTIMATION ❑ SAMPLING DISTRIBUTIONS AND STANDARD ERROR ❑ INTERVAL ESTIMATION FOR THE POPULATION MEAN ❑ LEVELS OF CONFIDENCE ❑ SAMPLING ERROR AND SAMPLE SIZE ❑ INTERVAL ESTIMATION OF THE DIFFERENCE BETWEEN TWO MEANS; MATCHED AND UNMATCHED SAMPLES ❑

FROM LITTLE TO LARGE: FROM SAMPLE STATISTICS TO POPULATION PARAMETERS

We call the characteristics of the population we are studying, the *population parameters*. Examples are the mean age of *all* females with the HIV virus; the proportion of *all* prescriptions which are for antidepressant drugs; the standard deviation in the time spent by *all* patients waiting for hernia operations. The values of population parameters are *fixed* and define the population. The thing is, we don't know what the value of any population parameter is because it is invariably impossible to identify (never mind find the time and resources necessary to study) every member of a population. However, we can use the results we get from samples to guess or *estimate* their value. The sample values we use to do this are called *sample statistics*.

For example, suppose the population is defined as all females diagnosed as HIV positive in the UK. We want to know what the mean age of this population of females is, i.e. we want the value of the population parameter, *mean age*. Suppose we take a representative sample, say of 1000 women, record the age of each, and calculate the sample mean age to be 20.8 years. This sample mean age is the *sample statistic* we are going to use to estimate the population mean age. Thus we could conclude or infer that the mean age of *all* UK HIV+ females is 20.8 years, because that's what the sample mean age is.

The most frequently used sample statistics and the population parameters they are used to estimate are shown below, together with the symbols commonly used to denote them (although there is no generally accepted symbol for either the sample median or the population median). Note that population parameters usually have Greek symbols, while sample statistics have English symbols.

- The sample mean ($\bar{x}$) is used to estimate the population mean μ, pronounced "mew".
- The sample proportion (p) is used to estimate the population proportion, π.

- The sample median (m) is used to estimate the population median, M.
- The sample standard deviation (s) is used to estimate the population standard deviation σ, "sigma".
- The sample interquartile range (iqr) is used to estimate the population interquartile range, IQR.

MEMO

Estimation is the process of using the value of some sample statistic to estimate the value of a corresponding population parameter.

Common sense tells us that a sample is never going to have *exactly* the same features and characteristics of the population from which it was taken. So we can't expect the sample mean to be *exactly* the same as the population mean (except by rare coincidence). A crucial question then immediately arises, "How close to the population mean age is this sample mean age of 20.8 years likely to be?" The answer is that it depends on how similar the sample is to the population. Although we know that a sample won't be *identical* to a population, if carefully taken (for example if we use random sampling), the sample should be a *reasonably close* representation of the population, and so the sample mean and population mean should not be too far apart.

Thus the process of statistical inference has a degree of *uncertainty* associated with it, which arises because we're never quite sure how representative of the population the sample is. We may be very unlucky and get a sample which (even though we were very careful and took a proper random sample) contains a disproportionately large number of individuals who are untypical of the population as a whole. In practice in the health and human sciences we often have to work with samples which are far from being random, for example samples which consist of *presenting* individuals. If we are going to have to deal with uncertainty we will need to know a little bit about probability.

MEMO

Because a sample will not be *exactly* the same as its parent population, a sample statistic will not be exactly the same as the corresponding population parameter. The process of statistical inference thus has a degree of uncertainty about it.

CHANCE WOULD BE A FINE THING: PROBABILITY THEORY

As we will see shortly the notion of probability will help us decide how accurate any given sample statistic is as an estimator of a population parameter. Although probability theory could occupy a book much bigger than this on its own, we will need only the barest essentials.

We start by noting that whenever we take a sample from a population we are in a sense carrying out an *experiment*. For example, suppose that as part of a cross-sectional study we are interviewing a sample of mothers of children who have died from sudden infant death syndrome (SIDS). One of the questions we might ask the mother is, "How did you feed your baby?" and ask her to choose from three possible responses, "Fully breast fed", or "Mixed breast/bottle", or "Fully bottle fed". Every time we ask this question it's like conducting an experiment, the outcome of which is one of these three responses or outcomes. Suppose we are interested in the probability of getting the response "Fully breast fed" from the next mother we interview. How can we determine what this probability is?

Calculating Probabilities

Regardless of how we calculate probabilities, there are two conditions that must be satisfied. The first condition is that the probability of getting any one of all the possible outcomes from an "experiment" (such as responses to a questionnaire) must lie in the range 0, meaning impossible – will never happen, to 1, meaning certain – will definitely happen. The second probability condition is that the sum of the probabilities of *all* possible outcomes from an experiment must equal 1. That is, the probability of getting the response "Fully breast fed" plus the probability of the response "Mixed breast/bottle" plus the probability of the response "Fully bottle fed" must equal 1.

There are three ways of calculating the probability of any outcome. The first is known as the *classical* method, so called because it stems from the work of early researchers in probability theory. This approach only works with experiments where all outcomes have an equal chance of happening (like rolling a dice or tossing a coin) and has little relevance in any practical studies in the human sciences. The second approach is much more useful and is known as the *relative frequency* method. This method depends on the existence of data, in the form of a frequency distribution, on identical or very similar experiments in the past. This information is then applied to a current experiment.

The third method, known as the *subjective method,* depends on a subjective judgement of the likely probabilities by the researcher concerned, using every scrap of relevant information that she or he can find, maybe after consulting with colleagues and with perhaps a touch of intuition and gut feeling. This is the only feasible approach where suitable frequency data do not exist (for example, the probability that a brand new drug for the treatment of AIDS will be effective).

An Example from Practice

Researchers[1] investigating the contribution, if any, of feeding method on sudden infant death syndrome obtained the data shown in Table 2.1 on 98 babies.

Using the relative frequency method, the probabilities of receiving each of the three responses are:

probability of fully breast fed = 0.17
probability of mixed breast/bottle fed = 0.40
probability of fully bottle fed = 0.43

Table 2.1: Number and percentage of babies dying from sudden infant death syndrome, by type of feeding

Type of feed	Number dying of SIDS	Percentage dying of SIDS
Fully breast fed	17	17
Mixed breast/bottle fed	39	40
Fully bottle fed	42	43

Notice that each of these probabilities is between 0 and 1, and together they add up to 1.

Although we have scarcely brushed the subject of probability theory, we have done enough for now. The main thing is that values of probability close to 1 mean that the outcome in question is quite likely to happen; and values close to 0 mean that the outcome is unlikely to happen.

MEMO

The probability of a particular outcome happening varies from 0 to 1. A probability of 0 means the outcome will definitely not happen. A probability of 1 means the outcome is certain to happen. The closer the probability is to 1 the more likely is it that the outcome will happen.

ESTIMATING THE POPULATION MEAN μ

Researchers conducting a randomised control trial[2] to investigate the short-term efficacy of a self-directed treatment manual for bulimia nervosa, measured the basic characteristics of a presenting sample of 81 consecutive referrals to a clinic for patients with bulimia nervosa. Of these, 41 were assigned to the "manual" group, 21 to a group to receive cognitive behavioural therapy and 19 to a waiting list group. The sample mean age $\bar{x}$ of the manual group was measured as 25.7 years with a sample standard deviation s of 5.8 years.

We can use this sample mean value of 25.7 years as a *point estimate* of the population mean age μ (the population here being defined as *all* such bulimics of whom this particular sample is representative). We refer to the sample mean as a *point* estimate because it is the *single* best guess we could make of the mean age of this population. If we had to put our last penny on the most likely value this is the one we'd pick. We are going to make two assumptions; first, that the sample was a random sample from the population, and second, that the distribution of age in this population is Normal. I will have more to say about these assumptions later.

Common sense tells us that *no* sample will ever be an *exact* representation of the population from which it is drawn so we cannot expect the population mean age to be *exactly* the same as the sample mean age. However, if the sample is a *random* sample then we can be *reasonably confident* that the sample mean and the population mean won't be all that far apart (after all, the whole purpose of random sampling is to produce samples which are good approximations to their populations). The crucial question is how far apart are they likely to be? Is the sample mean of 25.7 years within 6 months of the population mean, within a year, within 5 years? Without this extra information on its likely accuracy the sample mean (or any other point estimate) is only of limited value*, although its a lot better than nothing!

MEMO

A point estimator (for example a sample mean) is the single best guess we can make as to the value of the population parameter (for example a population mean). But it tells us nothing about how accurate an estimator it is.

How then can we find out how close to a population mean any particular sample mean is likely to be?

Suppose you and your friend Vlad are thinking of a holiday to the Seychelles. If your travel agent told you that the average mid-day temperature was likely to be 82°F, plus or minus 5°F, you would have a pretty good idea what to expect. You could be *reasonably confident* that the mid-day temperature most days was likely to be between 77°F and 87°F, but you would also know that on some days it might be a bit cooler, on others a bit warmer. What your travel agent has supplied you with is known to statisticians as an *interval estimate* of the average mid-day temperature. The interval in this case is from 77°F to 87°F, and the value of 82°F is the *point* estimate of average mid-day temperature (and it lies in the middle of the interval).

Suppose now that you press your travel agent a little further and ask them how much confidence do they have in these figures. They might reply, "fairly

* It's worth reiterating that we can never actually know the value of any population parameter including the population mean, because, for example, there is no way we could identify, locate and interview every person currently suffering from bulimia nervosa.

confident", or "very confident", or if they were numerically minded, "90 per cent confident", or "99 per cent confident". In the former case you might deduce that on nine days out of ten the temperature was likely to be between 77°F and 87°F, and on the remaining day it could be cooler or warmer than this. In the latter case you could expect the temperature would be between 77°F and 87°F almost all of the time (99 days out of 100) and only very occasionally outside this interval.

When a degree of confidence is attached to an interval estimate, the estimate is usually referred to as a *confidence interval estimate,* or just *confidence interval* for short. The point of all this is that we can use exactly the same sort of approach when we want to know how close a sample mean is to a population mean. In other words, we calculate confidence interval estimates of the population mean, usually at the 90, 95 or 99% levels.

The discussion in the next few pages may contain more algebra or difficult ideas than you feel you can cope with at the moment. If so my advice would be to skim read this section until you get to "Using a Computer for Confidence Interval Estimates of the Population Mean" on page 22. You can always come back and read these pages more carefully at a later time.

Interval Estimation of μ: the Large Sample Case ($n > 30$)

To return to the discussion on calculating confidence intervals, we can now indulge in what Einstein and many other theoretical physicists call a "thought experiment". Let's go back to our bulimia nervosa example where the sample of 41 subjects had a sample mean age $\bar{x}$ = 25.7 years. Imagine now that we take another random sample of 41 individuals from the same population, and then another sample, and then another, and another, and so on until we have taken *all* the different samples from this population that it is possible to take. Depending on the size of the population it might be possible to take hundreds, thousands, or even millions of different samples of size 41. For example, with a population of only 100, it is possible to take 203×10^{26} different samples of size 41 (that's 203 with 26 zeros after it!) which is why it is referred to as a thought experiment, because in practice we couldn't actually carry it out.

Now imagine that for each one of these samples we calculate the sample mean age. Each of these sample mean age values will be different because no two samples will be exactly the same, i.e. will not contain exactly the same 41 individuals from the population. Suppose now that we arrange all of these different sample means into a frequency distribution*. Statisticians refer to a frequency distribution of all possible sample means as a *sampling distribution* because it comes about from taking samples.

MEMO

The frequency or probability distribution of all possible sample means is called the sampling distribution of the sample mean, or just the sampling distribution of the mean.

* More properly known as a probability distribution.

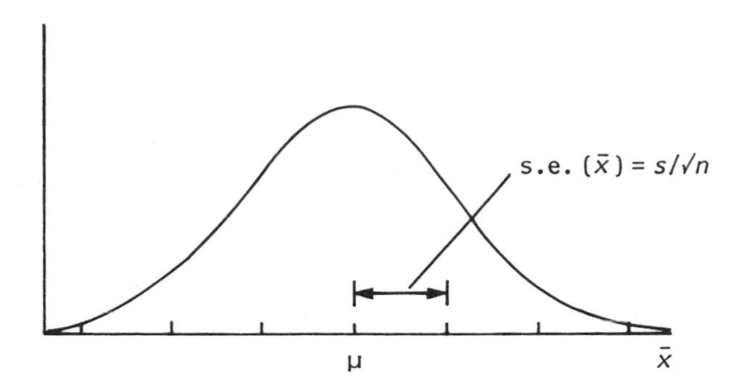

Figure 2.1: The sampling distribution of sample means

The reason for doing all this is that it can be shown that any sampling distribution made up of all possible sample means has the following crucially important properties (illustrated in Figure 2.1):

- It has a mean equal to the population mean μ.
- It has a Normal shape.
- It has a standard deviation equal to the standard deviation of the *population* divided by the square-root of the sample size, i.e. a standard deviation equal to $\sigma/\sqrt{n}$.

The first property implies that *on the whole* the sample mean is an accurate (statisticians use the word *unbiased*) estimator of the population mean. In other words the mean of all possible sample means is equal to the population mean μ.

The second property means that if the population from which the sample is taken is Normally distributed then the sampling distribution of sample means will also be Normally distributed regardless of sample size. Not only that, even if the population is *not* Normally distributed, the sampling distribution of sample means has the rather remarkable property that it will *still* be *approximately* Normally distributed. Moreover, the approximation is very close except for extremely small samples. This property is due to what is called the *Central Limit Theorem*.

MEMO

The sampling distribution of the mean is normally distributed with a mean equal to the population mean μ and a standard deviation (called the *standard error*) approximately equal to the sample s.d. divided by the square-root of the sample size, i.e. $s/\sqrt{n}$.

The third property implies that as the sample size n gets larger the spread of the sampling distribution of sample means gets smaller, i.e. the range of possible sample mean values gets smaller. As we will see, the second and third properties together mean that we can use the standard Normal distribution to construct confidence interval estimates of the population mean μ. We need to note two things before we see how to do that.

First, the standard deviation of the sample mean has come to be known as the *standard error* and usually written s.e.($\bar{x}$). If this is confusing, simply translate in your mind "standard error" whenever it occurs into "standard deviation of all possible sample means".

The second point is that we don't know the value of the population standard deviation σ (recall that we don't know the value of any population parameter because we can't study whole populations). We get round this by using the *sample* standard deviation s as a proxy for the unknown σ. In other words the standard error of the sample mean becomes $s/\sqrt{n}$ (the error involved in using s as a substitute for σ is acceptably small if n is 30 or more; we will examine the situation when n is less than 30 later on).

Let us see how these ideas work in practice. If we go back to the bulimics age example, with a sample standard deviation $s = 5.8$ years and a sample size $n = 41$, then the standard error of the sample mean age is therefore:

$$\text{s.e.}(\bar{x}) = s/\sqrt{n} = 5.8/\sqrt{41} = 5.8/6.403 = 0.906 \text{ years.}$$

MEMO

The standard deviation is a measure of the spread of the values taken by some variable in a particular *single* sample.

The standard error is a measure of the spread of the values of the means of *all* the samples it is possible to take from some population.

So the distribution of all possible sample means has a Normal shape, is centred around the population mean age (whose value we don't know), and has a standard deviation or standard error of 0.906 years. Referring back to the discussion of the area properties of the standard Normal distribution described in *Statistics from Scratch*, it follows that there is a 95% chance that our sample mean of 25.7 years will lie no more than *about* two standard deviations, or *about two standard errors* (to use the more conventional jargon) from the population mean μ. That is:

$$25.7 \text{ lies within } \mu \pm 2 \times 0.906$$

$$\text{or} \quad 25.7 \text{ lies within } \mu \pm 1.812.$$

Clearly it would be better to provide an *exact* rather than an approximate 95% confidence interval estimate of the population mean age, which we can do by consulting the table of the standard Normal distribution values in Appendix A1. Note that 95% *either* side of the mean is the same as 47.5% *each* side of the mean, so we need the value of z corresponding to a proportion of 0.475. This turns out to be $z = 1.96$ (check to make sure you can get this value). In other words 95% of sample means will lie within $1.96 \times \text{s.e.}(\bar{x})$ or $1.96 \times s/\sqrt{n}$ of the mean. Thus we can be 95% confident that:

$$25.7 \text{ lies within } \mu \pm 1.96 \times 0.906$$

$$\text{i.e.} \quad 25.7 \text{ lies within } \mu \pm 1.776.$$

This is illustrated in Figure 2.2, which shows how all possible values of the sample mean $\bar{x}$ are distributed around the population mean μ (even though the value of this is not known).

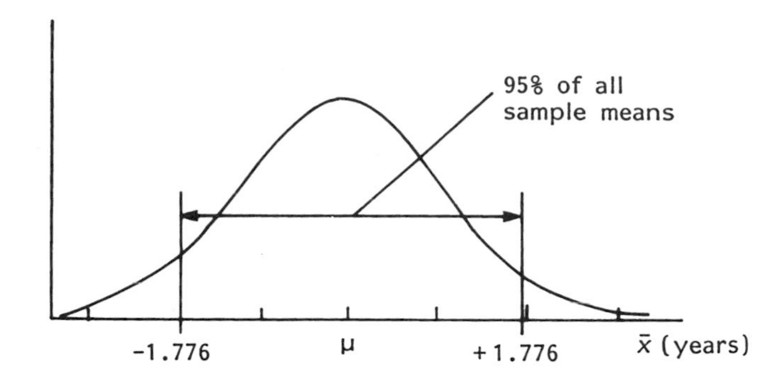

Figure 2.2: Distribution of sample mean ages around the (unknown) population mean—95% of all samples means lie plus or minus 1.776 years from μ

But we want to estimate the value of μ not $\bar{x}$, so we need to turn the above expression round. If we do this we can state that we can be 95% confident that:

μ lies within 25.7 years ± 1.776 years.

In other words, we can be 95% confident that the population mean age lies somewhere between 23.924 years and 27.476 years. Its more likely to lie towards the centre of the interval because of the humped shape of the Normal distribution (loosely speaking the higher the curve the higher the probability that the population mean will lie under it). A confidence interval is usually written in the form (23.924, 27.476) or (23.924 to 27.476) with the level of confidence stated, e.g. 95% or 99%.

There is one important aspect of this resulting confidence interval that is tied to the discussion later in this book on hypothesis testing. Suppose we had believed, before calculating this confidence interval, that the mean age μ of the bulimic patients in the population from which this sample was taken was 21 years. The fact that the confidence interval, from 23.924 to 27.476, does *not include* the value 21, means that we would have to conclude that our previously held view about the mean age of bulimic patients was mistaken. In other words we would have to *reject* our hypothesis that $\mu = 21$. On the other hand, if we had previously believed that the mean age of bulimics in this population was, say, 26 years, then this confidence interval would *support* this belief or hypothesis, because the value 26 does lie in the interval.

The important point is that we can use a confidence interval not only to inform us about the likely value of a population parameter, but also to test a belief or hypothesis we might hold about the value of that population parameter. If the confidence interval includes the believed or hypothesised value we can conclude, with a 95 or 99% certainty anyway, that our hypothesis is true. If the confidence interval does not include the value we previously believed the population parameter to be equal to, then we must reject that belief or hypothesis. So, not only does a confidence interval *estimate* it also *tests*!

In general, for any given sample mean $\bar{x}$ and sample standard deviation s, the 95% confidence interval for the population mean μ is:

$$\bar{x} \pm 1.96 \times \text{s.e.}(\bar{x})$$

$$\text{or} \quad \bar{x} \pm 1.96 \times s/\sqrt{n}.$$

MEMO

If you believe that a population parameter has a certain value and the confidence interval turns out to include that value, then you can be 95% or 99% confident that the population parameter is equal to that value. If the confidence interval does *not* include that value, then you can be 95% or 99% confident that the population parameter is not equal to that value. In this way a confidence interval can act as a way of testing beliefs about values of population parameters.

So 95% of all possible sample means will lie within $\pm$ 1.96 $\times$ s.e.($\bar{x}$) of the population mean μ. Another way of interpreting this result is that the interval from $\bar{x} - 1.96 \times s/\sqrt{n}$ to $\bar{x} + 1.96 \times s/\sqrt{n}$ will contain the population mean μ. That is, if we took 100 such samples, in about 95 of them the interval would contain the population mean, but about five would not. It's important to realise that a different sample to this would produce a different sample mean and consequently a different interval (although of similar width). We can't be absolutely *certain* that any particular sample we take gives us an interval that definitely *does* contain μ, we have to put our trust in probability theory and hope that we've got one of the 95% of the samples where it does and not one of the 5% of samples where it doesn't!

Confidence levels at the 95% level are much the commonest, but referring to Table A1, the 90% confidence interval is:

$$\bar{x} \pm 1.64 \times \text{s.e.}(\bar{x})$$

$$\text{or} \quad \bar{x} \pm 1.64 \times s/\sqrt{n}.$$

Similarly, the 99% confidence interval is:

$$\bar{x} \pm 2.58 \times \text{s.e.}(\bar{x})$$

$$\text{or} \quad \bar{x} \pm 2.58 \times s/\sqrt{n}.$$

The 99% confidence interval for population mean age works out to be (23.362, 28.037), which is a wider interval than that for a 95% level of confidence. Not surprisingly, the more confident we want to be that the interval contains the population parameter, the wider the interval has to be. If we wanted to be 100% confident we would have to have an interval going from minus infinity to plus infinity!

The general expression for a confidence interval for the population mean can thus be written:

$$\bar{x} \pm z \times \text{s.e.}(\bar{x})$$

$$\text{or} \quad \bar{x} \pm z \times s/\sqrt{n}$$

where z = 1.64 for a 90% confidence interval estimate, 1.96 for a 95% confidence interval estimate, or 2.58 for a 99% confidence interval estimate.

We can summarise the process of calculating a confidence estimate of the population mean μ as follows:

- *Step 1:* Calculate the sample mean $\bar{x}$.
- *Step 2:* Calculate the sample standard deviation s.
- *Step 3:* Calculate $s/\sqrt{n}$, the standard error of the $\bar{x}$, s.e.($\bar{x}$), by dividing s by root n.
- *Step 4:* Choose the desired level of confidence and the corresponding value of z (1.64, 1.96 or 2.58).
- *Step 5:* Multiply z by the value obtained in Step 3.
- *Step 6:* The confidence interval is then $\bar{x}$ plus and minus the value obtained in Step 5.

MEMO

The higher the confidence level the wider and therefore less precise is the confidence interval.

An Example from Practice

In a randomised trial to evaluate the effectiveness of a supported education programme for asthma patients,[3] a total of 801 patients were allocated either to an enhanced support treatment group (n = 397 patients) or the control group (n = 404). The main outcome measures included: number of hospital admissions; number of consultations for asthma with GPs; number of bronchodilators prescribed; number of days of restricted activities; and number of disturbed nights.

Table 2.2 shows the point estimates, followed in brackets by the 95% confidence intervals, for the population mean numbers in each category, for both the treatment and control groups, after one year.

Table 2.2: Point and 95% confidence interval estimates for numbers in various categories for treatment (enhanced support) and control groups in a randomised trial of the efficiency of enhanced support for asthma patients

CLINICAL OUTCOME	CONTROL GROUP	TREATMENT GROUP
Bronchodilators prescribed	10.5 (9.6, 11.5)	9.3 (8.1, 10.5)
GP consultations	2.6 (2.3, 2.9)	2.6 (2.2, 3.0)
Hospital admissions	0.19 (0.15, 0.24)	0.09 (0.06, 0.14)
Disturbed nights per week	2.1 (1.9, 2.3)	1.6 (1.4, 1.9)
Days of restricted activity per week	6.7 (5.3, 8.5)	4.2 (2.6, 6.8)

Thus the point estimate for the number of bronchodilators prescribed in the population from which this sample was taken was 10.5 for the control group and 9.3 for the enhanced support (treatment) group. This appears to demonstrate that the supported group experienced less severe asthma during the study period. However, the corresponding confidence intervals were (9.6 to 11.5) for the control group, and (8.1 to 10.5) for the supported group. The fact that these intervals overlap *suggests* that there was no statistically significant difference between the two populations in terms of the number of bronchodilators prescribed.

However, we would need a more rigorous analysis of the difference between the means of the two groups before we could be certain of this; we will examine such a procedure later in this chapter. In fact the only significant differences in the two groups were between the numbers of hospital admissions and the numbers of nights of disturbed sleep.

Sampling Error and Confidence Level

The difference between the value of the population mean μ and the sample mean $\bar{x}$ is known as the *sampling error*. In the bulimia nervosa example above we showed that 95% of all possible sample means lie within plus or minus $1.96 \times s/\sqrt{n} = 1.776$ years of the population mean age μ. In other words, 95% of all sample means give us a sampling error of *no more* than 1.776 years. Figure 2.3 shows the location of all the sample means that give a sampling error of 1.776 years or less.

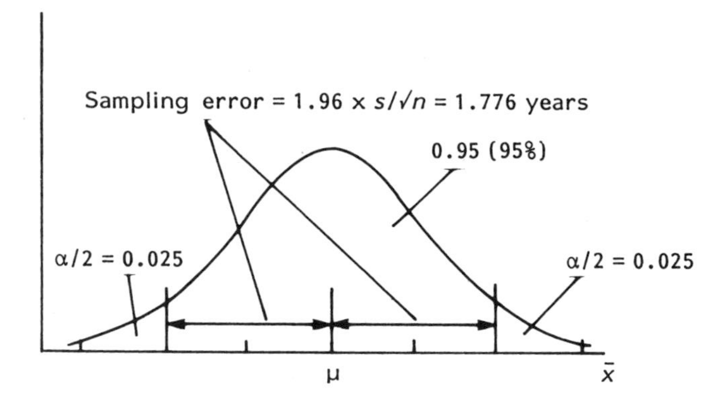

Figure 2.3: Sampling error

> **MEMO**
>
> The difference between the true value of a population parameter and the sample statistic is known as the sampling error.

In probability terms, there is a *probability* of 0.95 that the sampling error will be no more than 1.776 years. However, as we noted above, it's possible for a sample mean to fall outside the area shown in Figure 2.3 and produce an error in excess of 1.776 years, but the probability of this happening is only $1 - 0.95 = 0.05$. The probability of a particular sample mean being further away from the true population mean than $\pm z \times s/\sqrt{n}$ is denoted α (alpha). That is, with an α level of 0.05 we run the risk of being wrong (the interval not containing μ) five times in 100 samples. Since, in these circumstances, the interval may either be below μ or above it, the probability of one or the other of these happening is $\alpha/2$, i.e. $0.05/2 = 0.025$.

Using the same notation, the probability of being correct (the interval does contain μ) is therefore $(1 - \alpha)$. The term $(1 - \alpha)$ is known as the *confidence level* of the interval estimate*. These ideas are illustrated in Figure 2.4.

---

* The confidence level is usually expressed in percentage terms, i.e. when $(1 - \alpha) = 0.95$, the confidence level is 95%.

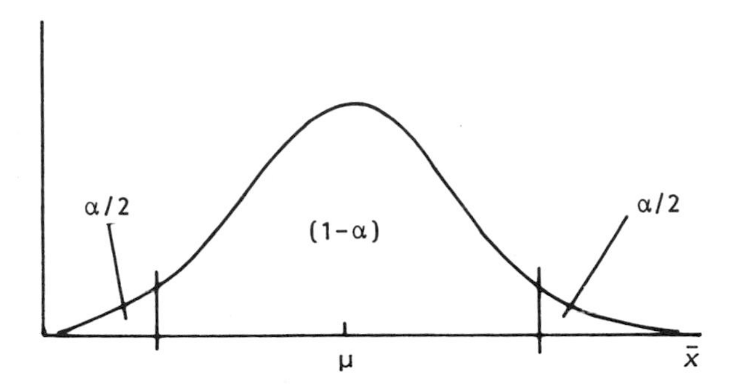

Figure 2.4: Sampling error and areas under the Normal sampling distribution of $\bar{x}$

> **MEMO**
>
> The probability that the confidence interval will not contain the value of the population mean μ is denoted α. The probability that the confidence interval will contain μ is equal to $(1 - \alpha)$ and is known as the confidence level.

Sample Size

Suppose the researchers were unhappy about a sampling error as large as 1.776 years and wanted to reduce it to say one year. The only way to reduce the size of the maximum likely sampling error is to increase the size of the sample. We saw above that, for a 95% confidence interval:

$$\text{Maximum likely sampling error} = 1.96 \times s/\sqrt{n}.$$

If we denote the maximum likely sampling error as E, then:

$$E = 1.96 \times s/\sqrt{n}$$

and if we rearrange this equation we get:

$$\sqrt{n} = \frac{1.96 \times s}{E}.$$

Squaring both sides gives:

$$n = \frac{1.96^2 \times s^2}{E^2}$$

or more generally:

$$n = \frac{z^2 \times s^2}{E^2}$$

with the appropriate value of z. This expression can be used to determine the minimum size of a sample to give any desired level of sampling error, provided that we know the value of the sample standard deviation s. The problem is we can't know the value of s until we have first taken the sample and we can't take the sample until we know the sample size! The only way round this dilemma is either to use information gleaned from similar previous studies, or to make an intelligent guess using all the information and experience to hand, or to conduct a small pilot study with a sample size chosen arbitrarily. This dilemma was faced by researchers investigating the effect of carbon dioxide on asthmatic patients: "It was difficult to estimate a required sample size since this is the first investigation of the potentiating effect of a gas on allergen response . . .".

For example, in studies of bulimia nervosa, experience might indicate that most bulimic patients are aged between 12 and 30, and their age is approximately Normally distributed within this range. This would mean that three standard deviations either side of the mean, i.e. six s.d.s in total, cover the range of 30 – 12 = 18 years. So 18 divided by six gives a sample s.d. of three years. This would provide a starting point for calculation of an appropriate sample size, perhaps for a pilot study, which could subsequently be modified in the light of the results obtained. If we have absolutely no idea as to the likely size of s we can set it equal to the sample mean $\bar{x}$, as a first step.

We can now return to our problem of finding a value for n which will reduce sampling error E to one year. Since we already have, from our previous study, a value for sample standard deviation s of 5.8, substitution into the above expression gives:

$$n = \frac{1.96^2 \times 5.8^2}{1^2} = 129.23.$$

So we would need to take a sample of size $n = 130$ to reduce maximum likely sampling error to one year. Note that we always have to round the value for n up to the next integer.

Estimating the Population Mean with Small Samples ($n \leq 30$)

Up to now I have said that it's okay to use the sample standard deviation s as a substitute for the unknown population standard deviation σ when calculating s.e.($\bar{x}$), the standard error of the sample mean $\bar{x}$. However, when sample sizes are small, i.e. less than 30, this is no longer appropriate, because the risk of s and σ differing significantly becomes too high to be acceptable. However, if we switch from the z distribution to another distribution, known as the t distribution, which takes this increased risk into account, then we can continue to use s as a proxy for σ. There is a price to pay. The resulting confidence interval estimates are wider and therefore less precise. The t distribution is wider and flatter than the z distribution, reflecting the increased risk of s and σ differing significantly. The two distributions are compared in Figure 2.5.

With small sample sizes, the confidence interval for the population mean age becomes:

$$\bar{x} \pm t \times \text{s.e.}(\bar{x})$$

$$\text{or} \quad \bar{x} \pm t \times s/\sqrt{n}.$$

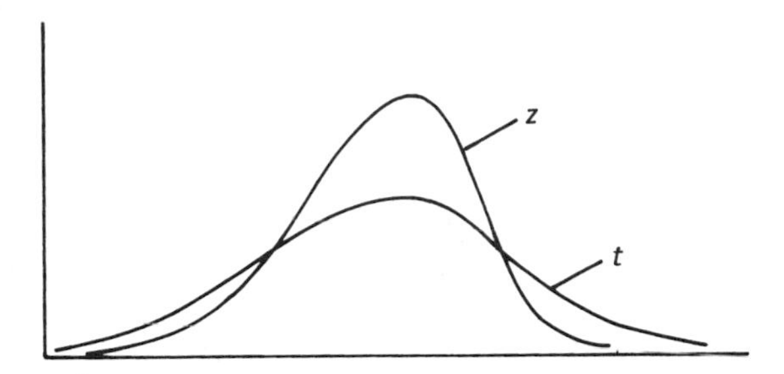

Figure 2.5: Comparison of z and t distributions

Values of the t distribution are given in Table A2 of the Appendix. Finding the appropriate value of t is only slightly more difficult than looking up a value of z, and we can illustrate the procedure by returning to the controlled trial bulimia example above. There were 21 patients assigned to the cognitive therapy group whose mean age was 26.0 years with a standard deviation of 6.6 years. The confidence interval estimate for this group is thus:

$$\bar{x} \pm t \times s/\sqrt{n}$$

$$\text{or} \quad 26.0 \pm t \times 6.6/\sqrt{21}.$$

A small section of the t table is shown in Table 2.3. The columns of a t table are usually labelled with the value of α, i.e. 0.01, 0.05, etc.

Table 2.3: Small section of the t table

DEGREES OF FREEDOM = $n - 1$	CONFIDENCE LEVELS		
	0.1 (90%)	**0.05 (95%)**	0.01 (99%)
20	1.725	**2.086**	2.845
21	**1.721**	**2.080**	**2.831**
22	1.717	**2.074**	2.819

To look up the appropriate value of t, we follow a two step process:

- *Step 1*. Choose the column corresponding to the required level of confidence, either in percentage or in probability terms, i.e. $(1 - \alpha)$. Suppose we want a 95% confidence level, i.e. $(1 - \alpha) = 0.95$. In which case we use the column headed 0.05. This column is shown bold in Table 2.3.
- *Step 2*. Choose the row corresponding to a value equal to the sample size minus 1, i.e. $(n - 1)$. This is known as the number of *degrees of freedom*, or *d.f.* (but there is no need to worry about what exactly this is). There are 21 patients in the cognitive behavioural therapy group, so $n = 21$ and $n - 1 = 20$. Thus we want row 20. This is shown bold in the table. The required value of t is found at the intersection of the 0.05 column and row = 20, i.e. 2.080. Notice that this value is a little larger than the 95% value of $z = 1.96$.

The 95% confidence interval for the population mean age of this group is given by:

$$\bar{x} \pm t \times s/\sqrt{n}$$

that is:

$$26.0 \pm 2.080 \times 6.6/\sqrt{21}$$

$$\text{or} \quad 26.0 \pm 2.080 \times 1.440.$$

(Compare this standard error of 1.440 to the standard error of 0.906 of the larger, $n = 41$, "manual" group.) The confidence interval for the population mean age of the cognitive therapy group is therefore:

$$26.0 \pm 2.996 \text{ years.}$$

Contrast the larger sampling error of this interval with that for the manual group of ± 1.776 years, a reflection of the smaller sample size and the consequent decrease in the precision of the estimate. So the 95% confidence interval estimate is from 23.004 years to 28.996 years, or (23.004, 28.996). The widths of such interval estimates are useful reminders of the imprecision of estimates.

As a final point, confidence interval estimates for the population mean are of course only appropriate when it has been decided that the mean itself is a suitable measure of location, i.e. when the variable concerned is metric and the distribution is reasonably symmetric or at least not markedly skewed. If the data are skewed or the variable is ordinal we might prefer the median as a measure of location and accordingly wish to calculate confidence intervals for the median rather than the mean. We will discuss this situation in Chapter 4. If the variable is nominal and we are dealing with the frequency of data in a number of categories we will need to consider confidence intervals for population proportions and possibly other approaches. This situation will be discussed in Chapter 3.

The method just described using the t distribution is based on the assumption that the population has a Normal distribution. Recall that methods based on such an assumption of Normality are known as *parametric* methods. However, when it is suspected, perhaps from examination of the shape of the sample data, that the population is *not* Normally distributed, or alternatively when no reliable information about its shape can be made because perhaps the sample size is too small to come to any reliable conclusion, then methods based on the t distribution are not appropriate. Non-parametric methods which do not rely on the shape of the population being Normal include calculation of confidence intervals for the median, and several tests of hypothesis discussed in Chapter 7.

Using a Computer for Confidence Interval Estimates of the Population Mean

Although the above discussion has been useful in enabling us to establish some of the important ideas underpinning interval estimation, it is becoming increasingly rare not to have access to one or more of the many statistical packages that will do the same calculations. In this section we are going to explore the use of some of these packages for calculating confidence interval estimates for the population mean. Let's start with the data in Table 2.4 which shows the ages of all persons

> **MEMO**
>
> Parametric statistical methods are based on the assumption that the distribution of the population is Normal. Non-parametric, or distribution-free methods, do not rely on this assumption.

who committed suicide in 1992, 1993 and 1994 (excluding December 1994), and who had had some form of contact with a particular mental health service in the 12 months prior to their suicide.

Table 2.4: Age distribution of persons committing suicide in 1992, 1993 and 1994 and who had had some form of contact with a mental health service in the 12-month period prior to their suicide

SUBJECT NUMBER	MALES (AGE)	FEMALES (AGE)	SUBJECT NUMBER	MALES (AGE)	SUBJECT NUMBER	MALES (AGE)
1	45	54	34	22	67	17
2	21	52	35	68	68	23
3	21	68	36	48	69	31
4	22	43	37	25	70	40
5	24	72	38	38	71	64
6	27	21	39	41	72	38
7	36	17	40	57	73	34
8	21	26	41	50	74	47
9	24	63	42	26	75	18
10	24	73	43	78	76	27
11	30	42	44	31	77	33
12	47	73	45	43	78	50
13	63	25	46	69	79	33
14	28	39	47	65	80	30
15	36	49	48	22	81	60
16	21	55	49	39	82	30
17	22	59	50	54	83	61
18	28	32	51	30	84	51
19	30	79	52	40	85	36
20	72	38	53	32	86	30
21	60		54	45	87	68
22	66		55	32	88	51
23	34		56	31	89	57
24	26		57	45	90	23
25	20		58	36	91	48
26	35		59	20	92	32
27	27		60	34	93	86
28	53		61	30	94	24
29	44		62	36	95	52
30	68		63	79	96	54
31	49		64	41	97	37
32	26		65	25	98	46
33	45		66	65	99	27

Using Minitab for Confidence Interval Estimates

To use Minitab the data are first entered into columns 1 and 2 (say) of the Minitab worksheet, and named "males" and "females"*. The following commands will produce the output shown in Figure 2.6:

Stat
  Basic Statistics
    1-Sample t
      Select "Males"
        Select "Females"
          ⊙ Confidence interval
            (Accept default level of 95.0 or type required value)
              OK

```
MTB > TInterval 95.0 ''males'' ''females''

            N      Mean      St Dev     SE Mean     95.0 Percent
                                                    C.I.

Males       99     39.90     16.09      1.62        (36.69, 43.11)
Females     20     49.00     18.94      4.24        (40.13, 57.87)

MTB > TInterval 99.0 ''males'' ''females''

            N      Mean      St Dev     SE Mean     99.0 Percent
                                                    C.I.

Males       99     39.90     16.09      1.62        (35.65, 44.15)
Females     20     49.00     18.94      4.24        (36.88, 61.12)
```

Figure 2.6: Minitab output for 95% and 99% confidence interval estimators of population mean age of male suicides

Minitab calculates the size of the sample (N), the sample mean (Mean), and sample standard deviation (St Dev), the standard error of the sample means (SE Mean) and the confidence interval for the population mean age.

The 95% interval estimate for men of (36.69 to 43.11), is noticeably narrower than that for women of (40.13 to 57.87). This is because of the much smaller number of women in the sample. The smaller the sample, the greater the possibility of sampling error and the wider the interval has to be to ensure the same degree of confidence that the resulting interval does contain the population mean μ. The 99% intervals are wider than those for 95% because of the increased level of confidence.

Using SPSS for Confidence Interval Estimates

The male suicide age data are entered into column 1 (say) of the SPSS datasheet. The following commands will produce the output shown in Figure 2.7:

---

* See *Statistics from Scratch* for a brief description of data entry methods in Minitab, SPSS, etc.

Statistics
  Summarize
    Explore
      Statistics
        Select males
          (Accept default confidence level of 95% or type new value)
            Continue
              OK

```
SPSS for MS WINDOWS Release 6.0

AGE

Valid cases: 99.0  Missing cases: 0  Percent missing: 0

Mean 39.8990  Std Err 1.6176  Min 17.0000  Skewness 0.7842
Median 36.0000  Variance 259.0305  Max 86.0000  SE Skew 0.2426
5% Trim 38.9955  Std Dev 16.0944  Range 69.0000  Kurtosis -0.1804
IQR 23.0000  SE Kurt 0.4806

95% CI for Mean (36.6890, 43.1090)
```

Figure 2.7: SPSS output for 95% confidence interval estimators of population mean age of male suicides

SPSS provides, in addition to the output measures from Minitab, the number and percentage of missing cases, the minimum and maximum values, the range and interquartile range, the median and variance, the 5% trimmed mean, and the skew and kurtosis coefficients together with their standard errors. A number of different plots and charts are also optionally available (e.g. boxplots, histograms), as well as several different transformations to Normal (e.g. square-root, log, inverse, etc.)*.

Using Excel for Confidence Interval Estimates

Excel will also produce a confidence interval for the population mean automatically when the **Descriptive Statistics** procedure is applied.

ESTIMATION OF MEANS WITH TWO SAMPLES

So far we've looked only at statistical procedures associated with data from *one* population. For example, in Figure 2.6 we used Minitab to calculate the mean age of a sample of male and female suicides, but separately. We made no attempt to use these two sample means to determine whether the mean age at which men commit suicide in this population was the *same* or *different* from that at which women commit suicide. The sample mean ages of 39.9 years for men and 49.0

* I haven't the space to discuss transforming data, but if the data are not Normally distributed and are not therefore amenable to a parametric procedure, including anything involving the t distribution, then transforming the data, say by taking logs, will often cause them to be more Normal in shape. You can then use a parametric procedure involving the t distribution.

years for women would seem to suggest, quite strongly perhaps, that women in this population are on average older than men when they commit suicide, but these differences may be due purely to chance. Moreover, the fact that the confidence intervals in Figure 2.6, (36.69 to 43.11) for men and (40.13 to 57.87) for women, overlap, might cause us to hesitate before coming to such a conclusion. What we need is a procedure which specifically addresses itself to a comparison between the two population means. It is to this that we now turn. Again you may feel a sudden urge to skim the algebra in the next few pages and come back to it at a future date. If so, go forward to "Using a Computer for Two-sample Interval Estimation" on page 30.

Let us denote the population mean ages of these men and women as μ_1 and μ_2. If there is no difference in these ages then $\mu_1 = \mu_2$, or $\mu_1 - \mu_2 = 0$. So what we want is an interval estimate of $(\mu_1 - \mu_2)$. If this interval includes 0, then we can assume that there is no significant difference in the two population mean ages. If the interval does not include 0, we can assume that there is a significant difference. Not surprisingly, we can use the difference in the two sample means $(\bar{x}_1 - \bar{x}_2)$ as an unbiased point estimator of this value. Figure 2.6 shows this difference to be $39.9 - 49.0 = -9.1$ years. Judging by this point estimate of the difference between $\mu_1 - \mu_2$, the age of female suicides is greater by nearly 10 years than that of men, but we need to calculate the appropriate confidence interval before we leap to any hasty conclusions.

The confidence interval for the difference in two population means is given by the expression:

$$(\bar{x}_1 - \bar{x}_2) \pm t \times \text{s.e.}(\bar{x}_1 - \bar{x}_2)$$

where $\text{s.e.}(\bar{x}_1 - \bar{x}_2)$ is the standard error of the sampling distribution of the difference in sample means. The value of t is taken from the t table (such as Table A2) using the row (i.e. the d.f.) with a value of $(n_1 + n_2 - 2)$. With $n_1 = 35$ and $n_2 = 25$, we look up the value of t in row $35 + 25 - 2 = 58$. However, we won't usually have to calculate the confidence interval by hand since there are plenty of computer packages that will calculate it together with the corresponding confidence intervals for us. Before we see some examples of these, we need to distinguish between two quite distinct two-sample situations: *first* where the two samples are *matched*, and *second* where they are *independent*.

Matched Samples

With two *matched* or *paired* samples, each individual in one group is paired with an individual in a second group so that each member of the pair share certain common attributes, e.g. sex, age, occupation, and so on. As we saw in Chapter 7 of *Statistics from Scratch*, matching usually takes place in case–control studies and in clinical trials to avoid the problem of confounding variables. Sometimes, rather than matching at the person-to-person level, *frequency matching* is used. In these circumstances, the same *proportions* of males and females, or of each age group, or each socio-economic group, or whatever, are chosen.

Another form of matching is where two measurements are taken on the same individual, also known as *self-matching*, or *before and after* matching. For example, individuals might have their blood pressure measured before and after undergoing some relaxation exercise. Matched samples are necessarily the same size.

With matched samples, if you have to calculate the confidence interval by hand, the first step is to calculate the difference d in each matched pair of sample scores, and then substitute into the following equation:

$$\bar{d} \pm t \times s_d / \sqrt{n}$$

where $\bar{d}$ is the mean of the differences in the two sets of sample scores, s_d is the standard deviation of these differences, and n is the number of matched pairs.

An Example from Practice

As an example consider the data in Table 2.5 which gives the bladder capacity of six women suffering from urinary incontinence, before and after treatment with a new electrical modality in a physiotherapy department.[4]

Table 2.5: Bladder capacity (ml) of six women with urinary incontinence before and after electrical treatment in a physiotherapy department

PATIENT	BEFORE (ml)	AFTER (ml)	d
1	149	189	−50
2	217	155	+62
3	193	283	−80
4	203	250	−47
5	296	450	−154
6	87	200	−113

From the data in the table the mean and standard deviation of the differences d are:

$$\bar{d} = -63.667 \text{ ml} \qquad s_d = 73.644 \text{ ml.}$$

The appropriate value of t is found in row 6 − 1 = 5, in the 95% (or 0.05) column of Table A2, i.e. $t = 2.571$. The 95% confidence interval is therefore:

$$-63.667 \pm 2.571 \times 73.644/\sqrt{6}$$

$$\text{or} \quad -63.667 \pm 77.30$$

which is −140.96 ml to −13.63 ml.

Since this interval does not include 0 we would be able to conclude that this new treatment was successful in significantly increasing bladder capacity. Of course, this is an extremely small sample with no guarantee that the distribution of bladder capacities in the population of similar women is Normally distributed (a condition for the use of the t distribution). In practice we would probably wish to consider a different approach, perhaps a non-parametric confidence interval for the difference between medians (discussed in Chapter 4).

Independent Samples

In the case of two *independent* samples of size n_1 and n_2, the selection of individuals for one sample is not at all influenced by the selection of the individuals for a second sample. For example, a random selection of patients from one health practice might be compared in terms of the number of times they visit their GP in a 12-month period with the number of visits in the same period made to their GP by patients randomly selected from a different health practice. Independent samples might be the same or different sizes. If it is ever necessary to calculate the confidence intervals by hand, we can use the following expression (those readers who feel that too much algebra is bad for their health might want to omit this section):

$$(\bar{x}_1 - \bar{x}_2) \pm t \times \text{s.e.}\ (\bar{x}_1 - \bar{x}_2)$$

where

$$\text{s.e.}(\bar{x}_1 - \bar{x}_2) = \sqrt{\frac{s_1^2}{n_1} + \frac{s_2^2}{n_2}}$$

and where s_1^2 and s_2^2 are the squares of the sample standard deviations of the first and second groups respectively.

If it can be assumed that the two population standard deviations are equal (or approximately so) then we can use what is called the *pooled variance* approach. In this case the standard error of the difference in means becomes:

$$\text{s.e.}(\bar{x}_1 - \bar{x}_2) = \sqrt{s_p^2 \left(\frac{1}{n_1} + \frac{1}{n_2}\right)}.$$

Here s_p^2 is the pooled variance and is given by:

$$s_p^2 = \frac{(n_1 - 1)s_1^2 + (n_2 - 1)s_2^2}{n_1 + n_2 - 2}.$$

Confidence intervals are narrower and therefore more precise if we can make a justifiable assumption of equal standard deviations and use the pooled variance approach. This is because we are putting more information, about the relative size of the two population standard deviations, into the calculation.

To see how this works, let's calculate the 95% confidence interval for the difference in the mean ages of male and female suicides (raw data in Table 2.4). From the Minitab output in Figure 2.6:

$$\bar{x}_{\text{males}} = 39.90 \text{ years and } s_{\text{males}} = 16.09$$

$$\bar{x}_{\text{females}} = 49.0 \text{ and } s_{\text{females}} = 18.94.$$

The standard deviation values are close enough for us to consider the pooled variance approach. The pooled variance is:

$$s_p^2 = \frac{(99 - 1)16.09^2 + (20 - 1)18.94^2}{99 + 20 - 2} = 275.10.$$

Therefore the standard error is:

$$\text{s.e.}(\bar{x}_1 - \bar{x}_2) = \sqrt{275.101 \left(\frac{1}{99} + \frac{1}{20}\right)} = 4.0662$$

and the 95% confidence interval for the difference in the mean ages of male and female suicides is:

$$(39.90 - 49.00) \pm t \times 4.0662.$$

To get the appropriate value for t, we need to look up the value in the 95% (or 0.050) column and the d.f. = $n_1 + n_2 - 2 = 99 + 20 - 2 = 117$th row of Table A2. This gives a value of $t = 1.98$. Substituting this into the above expression gives a confidence interval of:

$$(39.90 - 49.00) \pm t \times 4.0662$$

$$\text{or} \quad -9.1 \pm 1.98 \times 4.0662$$

$$\text{or} \quad -17.1511 \text{ to } -1.0489 \text{ years.}$$

Since this interval does not include 0, we can be 95% certain that there is a statistically significant difference between the ages at which men and women in these populations commit suicide, and this difference is somewhere between 1 year and 17 years.

An Example from Practice

In a controlled trial[5] to assess the effectiveness of health checks by nurses in reducing the risk factors for cardiovascular disease in patients from general practice, subjects were randomly assigned to the intervention group or the control group. The two groups were thus *independent*. (An example from practice for matched groups is given in the computer applications below.) Those in the intervention group were visited by nurses who offered counselling on known risk factors and negotiated priorities and targets for risk reduction. 95% confidence intervals for the differences in the mean values between the control and intervention groups are shown in Table 2.6 for a number of variables.

Table 2.6: 95% confidence intervals for the difference in the means of a number of variables between the control and intervention groups in a study of reduction in risk factors for cardiovascular disease

FACTOR	DIFFERENCE (CONTROL MINUS INTERVENTION); MEANS AND 95% CONFIDENCE INTERVAL	
	Men	*Women*
Total cholesterol (mmol/l)	0.06 (–0.03 to 0.15)	0.20 (0.11 to 0.29)
Systolic blood pressure (mmHg)	2.8 (1.3 to 4.4)	3.4 (2.0 to 4.8)
Diastolic blood pressure (mmHg)	1.8 (0.9 to 2.8)	1.7 (0.9 to 2.5)
Body mass index (kg/m²)	0.13 (–0.15 to 0.41)	0.18 (–0.14 to 0.50)
PUFA intake*	–0.5 (–0.7 to –0.4)	–0.5 (–0.6 to –0.4)

* Polyunsaturated fat intake score

For men, improvement over the control group appears to have taken place in systolic and diastolic blood pressure and in PUFA (polyunsaturated fats) intake. For cholesterol and body mass index both intervals include 0 indicating no significant difference in the respective means between the control and intervention groups. For women, improvement over the control group appears to have taken place in everything but body mass index whose interval includes 0. There is no harm in repeating this general and important conclusion. Any time that a confidence interval measuring the difference between two population parameter values includes the value 0, we can conclude that there is no statistically significant difference between them.

Using a Computer for Two-sample Interval Estimation

Using Minitab for Two Independent Samples

Minitab estimates the confidence interval for the difference in two means using the **two-sample t** program. For example, for a confidence interval for the difference in the mean ages of male and female suicides, the data are entered into columns 1 and 2 of the Minitab worksheet and named "males" and "females". The following commands will produce the output in Figure 2.8:

```
MTB > TwoSample 95.0 ''males'' ''females'';
SUBC > Alternative 0.

Two sample T-Test and Confidence Interval

Twosample T for males vs females

              N       Mean     St Dev    SE Mean

Males         99      39.9      16.1        1.6
Females       20      49.0      18.9        4.2

95% C.I. for mu males – mu females: (–18.5, 0.3)

T-Test mu males = mu females (vs not =): T = –2.01   P = 0.056   DF = 24
```

Figure 2.8: Minitab output from two-sample t showing confidence intervals for male and female suicide ages

Stat
- **Basic Statistics**
  - **2-Sample t**
    - **⊙ Samples in different columns**
      - **First**
        - **Select "males" for First box***
          - **Select "females"**
            - **(Accept default 95.0 or type in required value)**
              - **OK**

If you have good reasons to believe that the spread of ages (the standard deviation) in the population is approximately the same for males and females, click on the **Assume equal variances** button, otherwise leave this unchecked (this

* If all your observations are in one column, say males and females mixed together, you would need to set up a second "subscript" column with say a 1 on the same row as each male age value and a 2 on the same row as each female age value. In which case you would need to click on the "Samples in the same column" button and identify the subscript column.

assumption enables the "pooled variance" approach to be used). The output shown in Figure 2.8 is *without* the equal variances assumption.

The 95% confidence interval is (–18.5 to 0.3) years. This result means that we can be 95% confident that the difference in the population mean ages of male and female suicides, $(\mu_1 - \mu_2)$, is somewhere between –18.5 and +0.30 years. Since this interval includes the value 0 (which means that $(\mu_1 - \mu_2)$ could be 0), we have to conclude that, on the basis of this sample evidence, there is *no difference* in the mean ages at which men and women in these populations commit suicide.

Interestingly, the sample standard deviations for the ages of male and female suicides – which may be taken as a guide to the population standard deviations – are from Table 2.6, 16.09 and 18.94 years respectively. These values are quite similar and if we feel that this justifies clicking the **Equal variances** button, the confidence interval for the difference in means becomes (–17.2 to –1.0). This interval does *not* now include 0 and we could be 95% confident that there was a statistically significant difference in the population mean ages of male and female suicides.

MEMO

If a confidence interval for the difference between two population means includes the value 0, we can be 95 or 99% confident that there is no difference between the population means.

Using SPSS for Two Independent Samples

In SPSS the data for male and female age at suicide need to be entered into the *same* column (say column 1) of the SPSS datasheet (this column is known as the **Test Variable**). In column 2, type a 1 on each male row and a 2 on each female row (this gives what is known as the **Grouping Variable**). The following commands will produce the output shown in Figure 2.9:

Statistics
  Compare Means
    Independent Samples T Test
      Select c1 for Test Variable
        Select c2 for Grouping Variable
          Define Groups
            Type 1 in Group 1 box
              Type 2 in Group 2 box
                Continue
                  Options (to change default 95%)
                    Continue
                      OK

The SPSS output is somewhat more detailed than is that from Minitab. First, a point estimate of the difference in the mean ages of male and female suicides is given (–9.1010 years). Second, a test of whether male and female population variances are equal or not (Levene's test) is applied. As we will discover later in

```
t-tests for independent samples of VAR00001

                 Number
Variable        of Cases       Mean          SD         SE of Mean

AGE

VAR00001 1         99         39.8990      16.094         1.618
VAR00001 2         20         49.0000      18.940         4.235

Mean Difference = -9.1010

Levene's Test for Equality of Variances: F = 1.243  P = 0.267

t-test for Equality of Means              95%

Variances       t-        df       2-Tail      SE of    CI for
               value                Sig        Diff     Diff

Equal          -2.24    117        0.027      4.067    (-17.157, -1.045)
Unequal        -2.01     24.84     0.056      4.534    (-18.440, 0.238)
```

Figure 2.9: SPSS output for the 95% confidence interval for the difference in the mean ages of male and female suicides, with two independent samples

our discussion of hypothesis testing, the value of $p = 0.267$ indicates that the variances are most likely to be even. In any case, confidence intervals are calculated for both equal and unequal variances. When population variances are assumed equal the confidence interval, (–17.157, –1.045), does not include zero and so there is a probability of 0.95 that there is a significant difference between the ages at which men and women in this population commit suicide (this is the same interval we obtained when we calculated the interval by hand above). When the variances are not assumed to be equal, the confidence interval, (–18.440, 0.238), does include zero and we cannot therefore assume that there is a significant difference in these ages.

Using Minitab for Matched Samples

As an example of two matched samples consider the data in Table 2.7 on the duration of the first and fourth seizures for 40 patients undergoing ECT treatment. Since it is the *same* patient in each case who has the first and then the fourth seizure, the data are matched. Examination of the raw data gives the impression that in general the fourth seizure lasts a shorter time than does the first seizure. The sample mean seizure durations turn out to be 35.075 seconds and 30.250 seconds respectively. We want to determine whether we can put this difference down to chance, or if it is real and statistically significant.

To use Minitab to calculate a confidence interval for the difference in means of matched samples, the first and fourth seizure duration data are entered into columns 1 and 2, say, of the Minitab worksheet. We then first have to use the Minitab **let** command to subtract one column from the other. This means clicking in the Minitab sessions window (the top half of the screen) and typing:

Table 2.7: Raw data for the duration of seizure of patients having first and fourth treatments with ECT

PATIENT NUMBER	DURATION OF SEIZURE (seconds)		PATIENT NUMBER	DURATION OF SEIZURE (seconds)	
	First	*Fourth*		*First*	*Fourth*
1	45	40	21	90	50
2	37	40	22	30	20
3	30	21	23	15	40
4	40	20	24	40	42
5	60	45	25	42	50
6	30	17	26	30	10
7	45	28	27	20	30
8	30	35	28	35	35
9	30	35	29	25	25
10	35	35	30	25	30
11	64	30	31	15	10
12	32	21	32	30	40
13	25	30	33	30	20
14	50	30	34	45	25
15	33	36	35	15	35
16	25	30	36	45	20
17	50	25	37	40	40
18	25	30	38	30	25
19	20	30	39	15	20
20	50	40	40	30	25

let c3 = c1 – c2.

This puts the differences in column 3, which is then named "diff". The same effect could be achieved using **Calc, Mathematical Expression**, but for such a simple operation, typing the command is faster and easier. The following commands will produce the output shown in Figure 2.10:

Stat
  Basic Statistics
    1-Sample t
      Select "diff"
        (Accept or change the default 95%)
          OK

The 95% confidence interval (0.38, 9.27) does not include 0 so we can be 95% certain that the difference between the first minus the fourth seizure duration is significant. The first seizures do last longer, by somewhere between 0.38 and 9.27 seconds. However, the 99% confidence interval, (–1.13, 10.78), does include 0, and at this confidence level we cannot be sure that the difference between the two seizure durations is not due to chance alone.

Using SPSS for Matched Samples

The SPSS program does allow for a matched or paired confidence interval to be calculated directly. The data are entered into columns 1 and 2, say, of the SPSS

```
MTB > TInterval 95.0 ''diff''.

Confidence Intervals

Variable     N     Mean     St Dev     SE Mean       95.0% C.I.
diff        40     4.82     13.91       2.20       (0.38, 9.27)

MTB > TInterval 99.0 ''diff''.

Confidence Intervals

Variable     N     Mean     St Dev     SE Mean       99.0% C.I.
diff        40     4.82     13.91       2.20      (-1.13, 10.78)
```

Figure 2.10: Minitab output for 95% and 99% confidence intervals for the difference in the duration of the first and fourth ECT seizures, with matched samples (raw data in Table 2.7)

datasheet and the **Data, Define Variables** commands used to name the variables **First** and **Fourth**. The output from the SPSS program is shown in Figure 2.11, and is achieved by clicking on the command sequence:

Statistics
Compare Means
Paired-Samples T Test
Select "First"
Select "Fourth"
Options (to change default 95%)
Continue
OK

```
SPSS for MS WINDOWS Release 6.0

t-tests for paired samples

               Number          2-tail
Variable      of pairs   Corr    Sig      Mean        SD     SE of Mean
FIRST                                    35.0750    14.769      2.335
                 40     0.415   0.008
FOURTH                                   30.2500     9.724      1.537

Paired Differences

    Mean        SD       SE of Mean    t-value     df     2-tail Sig
   4.8250     13.910       2.199        2.19       39        .034

95% C.I.  (0.375, 9.275)
```

Figure 2.11: SPSS output for the 95% confidence interval for the difference in the duration of the first and fourth ECT seizures, with matched samples (raw data in Table 2.7)

This result confirms the Minitab result that at the 95% confidence level there is a significant difference in the duration of the first and fourth ECT seizures.

Using CIA for Matched or Independent Samples

CIA will also calculate confidence intervals for the difference in two means, for both matched and independent samples. Space considerations do not allow examples of this to be shown.

SUMMARY

In this chapter we have ventured beyond descriptions of samples into the realm of statistical inference, i.e. into exploring the characteristics of the population from which the sample was taken. Confidence interval estimation of the value of a population mean uses as its starting point knowledge of the sample mean and standard deviation of a sample taken from that population. This information is used to establish the characteristics of the sampling distribution of the sample mean. The sampling distribution of the appropriate sample statistic is the basis for almost all confidence interval estimation. The distribution of the sample mean turns out to be Normally distributed when the population is Normally distributed, and, through the mechanism of the Central Limit Theorem, approximately Normally distributed even if the population is not Normal.

The difference between the value of the population mean μ, which remains unknown, and any particular sample mean $\bar{x}$, is known as the sampling error. With a 95% confidence interval there is a probability of 0.95 that the sampling error will be no more than $\pm t \times s/\sqrt{n}$, i.e. that the population mean will lie somewhere in the interval $\bar{x} \pm t \times s/\sqrt{n}$. This leaves a probability of 0.05 that the interval will not contain μ. This probability is denoted α. The probability that the interval will contain μ is denoted $1 - \alpha$, and is known as the confidence level of the interval.

After starting the chapter looking at confidence intervals for a single population mean using a single sample, we ended by examining confidence intervals for the difference between two population means, first when the two samples are independent and second when they are matched.

In the next chapter I will describe methods for estimating the population proportion.

EXERCISES

2.1 (a) What is the process of statistical inference? (b) What are "sample statistics" and "population parameters"? What is the relationship between them?

2.2 Explain why a sample is never (except by rare coincidence) an exact representation of a population. What are the consequences for the process of statistical inference?

2.3 Suppose a population consists of only six values, 1, 2, 3, 4, 5 and 6.

(a) Calculate the population mean and standard deviation (normally unknown).
(b) Take all possible samples of size $n = 2$ from this population. Calculate the sample mean of each sample.

(c) Calculate the mean and standard error of the sample means. Compare with the population mean and standard deviation.
(d) Arrange the sample means into a frequency distribution and comment on its shape.
(e) Repeat the above calculations for all samples of size $n = 3$. Compare results with those obtained above.

2.4 Calculate the 95 and 99% confidence interval estimates for the population mean number of smoking co-workers of subjects in a study[6] into passive smoking and heart disease, using the data in Table 2.8. Comment on your results. What is the population referred to here?

Table 2.8: Number of smoking co-workers for a sample of 45 non-smoking females

SUBJECT	1	2	3	4	5	6	7	8	9	10	11	12	13	14	15
SMOKERS	2	0	5	15	1	8	5	0	0	9	10	4	4	2	3
SUBJECT	16	17	18	19	20	21	22	23	24	25	26	27	28	29	30
SMOKERS	5	1	14	7	9	2	3	0	0	5	8	1	6	0	6
SUBJECT	31	32	33	34	35	36	37	38	39	40	41	42	43	44	45
SMOKERS	0	4	7	5	2	2	6	3	1	0	1	1	3	2	0

2.5 The data in Table 2.9 show the waiting time (weeks) for independent samples of patients waiting to be admitted for a hernia repair in two different hospitals. Calculate a 95% confidence interval for the difference in mean waiting times.

Table 2.9: Time spent waiting by two independent groups of patients for admission for a hernia repair in two separate hospitals

PATIENT	HOSPITAL A	HOSPITAL B
1	24	32
2	28	30
3	19	25
4	20	18
5	31	27
6	25	26
7	26	28
8	14	32
9	23	30
10	29	20
11	26	25
12	11	40
13	16	29
14	24	12
15	35	30
16	22	19
17	35	35
18	18	29
19	37	15
20	25	20

2.6 Assume the data referred to in question 2.3 are for two matched groups of patients. Calculate a 95% confidence interval for the difference in mean waiting times.

2.7 In a cohort study into weight growth in infants born to mothers who smoked during pregnancy, comparisons were made with non-smoking mothers. Table 2.10 shows the mean weight at birth, in grams, at three months, and at six months, of babies born to mothers with different smoking habits during pregnancy, and weight differences and 95% confidence intervals for these differences compared with babies born to non-smoking mothers. (a) Interpret the results. (b) Calculate the standard error and hence the sample standard deviation in weight at birth for girl babies of mothers smoking ≥ 10 cigs/day.

Table 2.10: The mean weight at birth, in grams, at three months and at six months, of babies born to mothers with different smoking habits during pregnancy, and weight differences and 95% confidence intervals for these differences compared with babies born to non-smoking mothers

Mother's smoking habit	*At birth*			*At 6 months*		
	Number of babies	Weight (g)	95% confidence interval for difference	Number of babies	Weight (g)	95% confidence interval for difference
Girls:						
Non-smokers	4904	3220		4895	7462	
1–9 cigs/day	1072	3132	(–121 to –55)	1071	7471	(–47 to 65)
≥ 10 cigs/day	228	3052	(–234 to –102)	227	7434	(–141 to 85)
Boys:						
Non-smokers	5334	3373		5330	8038	
1–9 cigs/day	1204	3226	(–139 to –75)	1204	7974	(–118 to –10)
≥ 10 cigs/day	245	3126	(–312 to –181)	245	8014	(–136 to 88)

REFERENCES

1. Gilbert, R. E. *et al.* (1995) Bottle feeding and the sudden infant death syndrome. *BMJ*, **310**, 89.
2. Treasure, J. *et al.* (1994) First step in managing bulimia nervosa: controlled trial of therapeutic manual. *BMJ*, **308**, 686–9.
3. Osman, L. M. (1994) Reducing hospital admissions through computer supported education for asthma patients. *BMJ*, **308**, 568–71.
4. Cawley, D. M. and Hendriks, O. (1992) Evaluation of the Endomed CV 405 as a treatment for urinary incontinence. *Physiotherapy*, **78**, 495–7.
5. Imperial Cancer Research Fund OXCHECK Study Group (1994) Effectiveness of health checks conducted by nurses in primary care: results of the OXCHECK study after one year. *BMJ*, **308**, 308–12.
6. He, Y. *et al.* (1994) Passive smoking at work as a risk factor for coronary heart disease in Chinese women who never smoked. *BMJ*, **308**, 380–9.

3

ESTIMATING THE POPULATION PROPORTION

❑ THE SAMPLE PROPORTION AS AN UNBIASED ESTIMATOR OF POPULATION PROPORTION ❑ INTERVAL ESTIMATION OF π ❑ SMALL AND LARGE SAMPLE ESTIMATION OF π ❑ SAMPLE SIZE CALCULATIONS ❑ USING A COMPUTER TO ESTIMATE π ❑ CONFIDENCE INTERVALS FOR THE DIFFERENCE BETWEEN TWO POPULATION PROPORTIONS ❑

A FAIR SLICE OF THE PIE

In the health and human sciences we are frequently interested in determining the *proportion* of individuals or items in a population who possess some attribute or characteristic. Examples are the proportion of non-normal cervical smears in a given batch, the proportion of patients who fail to keep an outpatient appointment, the proportion of children who smoke, the proportion of ambulances which arrive within 10 minutes of an emergency call, and so on*. The population proportion is usually denoted as π, and the principles underlying the calculation of a confidence interval for π are similar to those for the mean described in the previous chapter.

THE SAMPLE PROPORTION

In the previous chapter we saw that the best point estimator of the population mean μ was the sample mean $\bar{x}$, and we were able to use $\bar{x}$ as the basis for an interval estimate of μ. Not surprisingly, the sample proportion p turns out to be the best point estimator of the population proportion π, and in the same way forms the basis of its interval estimate. The sample proportion is easily calculated. As an example, suppose that in a sample of $n = 120$ patients admitted to an A&E department after a road traffic accident, the number x with a blood alcohol level above the legal limit is 42. Then the sample proportion p is:

$$p = x/n = 42/120 = 0.35.$$

When it comes to determining the confidence interval estimates of the population proportion, we need to consider two separate cases, "small" and "large" samples.

---

* It should be noted that interval estimates for values which are expressed in *percentages* are also encompassed by the methods described here. Any such values should initially be divided by 100 to give proportions, before the methods are applied. The resulting interval estimates can then be multiplied by 100 at the end to give percentage interval estimates if required.

The rule of thumb is that if $n \times p$ and $n \times (1 - p)$ are *both* greater than 5, then the sample can be considered as "large". In the above A&E example, $n \times p = 120 \times 0.35 = 42$, and $n \times (1 - p) = 120 \times (1 - 0.35) = 78$, both of which are (considerably) larger than 5, so we would class this as a "large" sample case.

In contrast, suppose we had a sample of 12 liver transplant patients, three of whom had developed postoperative infections. The sample proportion $p = 3/12 = 0.25$, so $n \times p = 12 \times 0.25 = 3$ (which is less than 5), and $n \times (1 - p) = 12 \times (1 - 0.25) = 9$. Since one of these values is less than 5 this would be classed as a "small" sample. We are likely to get a small sample if either n is small and/or π, and hence p, is a long way from 0.5. I used quote marks around small and large above because the sample size n could still be relatively large and yet either $n \times p$ or $n \times (1 - p)$ could be less than 5 because p is a lot closer to 0 or 1 than to 0.5. I'll now discuss each case in turn.

MEMO

If either $n \times p$ and $n \times (1 - p)$ are smaller than 5 then we have a "small" sample.

Small Sample Interval Estimation of π

If we were to take all possible samples of some given size n from a population, calculate the sample proportion for each sample and arrange all of these sample proportions into a frequency distribution (more properly known as a probability distribution) we would have the sampling distribution of the sample proportion. With small sample sizes this distribution has what is known as a *binomial* distribution (in the same way as the sampling distribution of the sample mean has a Normal distribution). In these circumstances, calculation of the lower and upper limits of the interval estimate is rather complicated and involves use of the equation for cumulative binomial probabilities. However, and fortunately for us, these values have already been calculated[1] for sample sizes up to 100 for 90%, 95% and 99% confidence levels. These are reproduced in Table A3 of the Appendix for samples sizes up to 25 (readers who need the values for larger samples will have to refer to the original source).

Using this table is very straightforward and our liver transplant example provides an illustration. Table 3.1 shows the required section of Table A3. Suppose we want a 95% confidence interval for the proportion π of all such patients with a postoperative infection. In the example we have a sample size of $n = 12$ and the number with a postoperative infection is $x = 3$. So we need to be on row 3, and in the 95% column. The lower limit of the interval estimate is 0.0549, the upper limit 0.5719. So the 95% confidence interval for π is (0.0549, 0.5719) or (5.49%, 57.19%). With a 99% confidence level the lower limit of the estimate is 0.0303, the upper limit 0.6552, and the confidence interval is (0.0303, 0.6552), i.e. from 3.03% of patients to 65.52% of patients.

These very wide intervals provide a rather imprecise interval estimate of the population proportion of liver transplant patients who suffer postoperative infection. This is a reflection of the quite small sample size as well as a sample

Table 3.1: Small section of Table A3 showing the lower, p_l, and upper, p_u, limits for interval estimates of the population proportion for 90%, 95% and 99% confidence levels

			CONFIDENCE LEVEL					
			90%		95%		99%	
n	x	p	p_l	p_u	p_l	p_u	p_l	p_u
12	2	0.1667	0.0300	0.4376	0.0209	0.4841	0.0090	0.5729
	3	0.2500	0.0721	0.5266	0.0549	0.5719	0.0303	0.6552
	4	0.3333	0.1233	0.6087	0.0992	0.6511	0.0624	0.7275

proportion (and probably therefore a population proportion) which is some distance from 0.5. It's important to remember that the population we are talking about here is not all patients who have liver transplants, but only that population for whom this *presenting* sample is typical.

MEMO

With "small" samples, i.e. both $n \times p$ and $n \times (1 - p)$ less than 5, the sample proportion p has a binomial distribution. This is difficult to work with. In these circumstances it's much easier to use Table A3 to calculate confidence intervals for π.

Large Sample Estimation of π

As sample size increases, the shape of the binomial distribution gets more and more like the Normal distribution. When $n \times p$ and $n \times (1 - p)$ are *both* greater than 5 then the shape is close enough to Normal* for us to use the Normal distribution to construct approximate confidence intervals for the population proportion. In these circumstances the standard error of sample proportions is approximately:

$$\text{s.e.}(p) = \sqrt{\frac{p(1-p)}{n}}$$

and a confidence interval estimate of π is given using the expression:

$$p \pm z \times \text{s.e.}(p).$$

that is,

$$p \pm z \sqrt{\frac{p(1-p)}{n}}.$$

The value of z in this expression will equal 1.64, 1.96 or 2.58 for 90, 95 or 99% confidence levels respectively. The term $z\sqrt{\frac{p(1-p)}{n}}$ is the *sampling error.*

* The resulting distribution is known as the Normal *approximation* to the binomial.

The procedure for finding a confidence interval estimate of the population proportion can be summarised as follows:

- *Step 1:* Calculate the sample proportion p.
- *Step 2:* Subtract the value of p from 1, and multiply the result by p again.
- *Step 3:* Divide the value obtained in Step 2 by n and take the square-root of the result.
- *Step 4:* Choose a confidence level, 90, 95 or 99%, and get the corresponding value of z, 1.64, 1.96 or 2.58.
- *Step 5:* Multiply the value from Step 3 by the value of z from Step 4.
- *Step 6:* The confidence interval is then from p minus the value obtained in Step 5, to p plus the value from Step 5.

MEMO

With "large" samples, i.e. both $n \times p$ and $n \times (1 - p)$ greater than 5, the sample proportion p has an approximately Normal distribution. In these circumstances it is easiest to use z values from Table A1 to calculate confidence intervals for π.

An Example from Practice

In an investigation into the familiarity of school children with illicit drugs,[2] researchers in 1994 used a questionnaire to elicit responses from a sample of 392 school children aged 14 to 15. Those who responded positively to the question, "Do you personally know anyone taking drugs?" were additionally asked "What drugs did they take?" Of the 228 positive responders, 168 indicated they personally knew people who used cannabis, i.e. a figure of 74%.

The sample proportion is $p = 0.74$, and $n \times p = 168.72$ and $n \times (1 - p) = 59.28$, both of which > 5 so this is a large sample case. The standard error of sample proportions is:

$$\text{s.e.}(p) = \sqrt{\frac{p(1-p)}{n}} = \sqrt{\frac{0.74(1-0.74)}{228}} = 0.029.$$

The 95% confidence interval estimate is thus:

$$p \pm z \times \text{s.e.}(p)$$

$$\text{or} \quad 0.74 \pm 1.96 \times 0.029$$

$$\text{or} \quad 0.74 \pm 0.057.$$

So the 95% interval estimate of the unknown π is (0.683, 0.797) or (68.3%, 79.7%). In other words, there is a probability of 0.95 that the (unknown) proportion of children in this population who personally know someone who has used cannabis is somewhere between 68.3 and 79.7%. Note that the maximum likely sampling error is 0.057 (5.70%).

Choosing a Sample Size

Following a similar approach to that in Chapter 2 for the sampling error of the sample mean, the maximum likely sampling error for the sample proportion is: $\pm z \times$ s.e.(p). If we call this error E then:

$$E = z\sqrt{\frac{p(1-p)}{n}}\ .$$

Rearranging this equation for n, by first squaring both sides and then taking terms across, gives:

$$n = \frac{z^2 p(1-p)}{E^2}\ .$$

Suppose in the example on children's experience of illicit drugs we want to reduce sampling error from 5.7 to 5.0%, i.e. from 0.057 to 0.05. The required sample size is therefore:

$$n = \frac{z^2 p(1-p)}{E^2} = \frac{1.96^2 \times 0.74(1-0.74)}{0.05^2} = 295.6.$$

Thus the sample size will have to be increased from 228 to 296 to achieve a maximum sampling error of no more than 5%.

Using a Computer to Estimate the Population Proportion

As far as I am aware, neither Minitab, SPSS nor EPI allow *direct* calculation of the interval estimate of a population proportion (and although it wouldn't be too difficult to write a short macro to perform this calculation, I feel this is a little beyond the reach of this book). In fact the only package that I am familiar with that does include this facility is CIA. (This is not to say, of course, that there are other packages out there which do include this facility. I would be interested to hear from readers if this is so.)

Output from CIA showing the 95 and 99% confidence interval estimates for the population proportion of women (compared with men) who committed suicide, based on the data in Table 2.4, is shown in Figure 3.1. The commands necessary to produce this output are:

Chapter 4 – Proportions and their differences
1 : Single Sample

The screen then displays the message:

SAMPLE SIZE:
NUMBER WITH FEATURE:

The total sample size was 99 men plus 20 women, i.e. 119, so first the value 119 is typed, the Return key pressed, and then the value 20 typed and the Return key pressed. CIA then produces first the *exact* confidence interval using the binomial distribution and then asks if you want the approximate interval using the Normal distribution.

The point estimate (i.e. the sample proportion) is $p = 20/119 = 0.1681$ or 16.81%. As we see, the CIA program uses the Normal approximation to the binomial to give a 95% confidence interval estimate for the population proportion of women of (0.101 to 0.235) or (10.1% to 23.5%). The 99% confidence interval estimate is given as (0.0796 to 0.2570) or (7.96% to 25.7%).

```
SAMPLE SIZE: 119
NUMBER WITH FEATURE: 20
OBSERVED PROPORTION = 0.168
% CONFIDENCE REQUIRED: 95

------------------------------------------

NORMAL method:
Standard Error of Proportion = 0.0343   NORMAL value = 1.96

95% CONFIDENCE INTERVAL FOR PROPORTION IS:
0.101 TO 0.235

------------------------------------------

Another level of confidence (Y/N)?

SAMPLE SIZE: 119
NUMBER WITH FEATURE: 20
OBSERVED PROPORTION = 0.168
% CONFIDENCE REQUIRED: 99

------------------------------------------

NORMAL method:
Standard Error of Proportion = 0.0343   NORMAL Value = 2.58

99% CONFIDENCE INTERVAL FOR PROPORTION IS:
0.0796 TO 0.257

------------------------------------------

Another level of confidence (Y/N)?
```

Figure 3.1: Output of CIA for 95% and 99% confidence intervals in the proportion of women suicides (raw data in Table 2.4)

ESTIMATING THE DIFFERENCE BETWEEN TWO PROPORTIONS

We will often be interested in estimating the difference between two proportions. For example, in the study reported above on school children's knowledge of illicit drugs we may want to know whether the proportion of girls who knew someone who used such drugs was the same as the proportion of boys. In these circumstances we want to form an interval estimate not of π, but of $(\pi_1 - \pi_2)$, where π_1 is the proportion in one population (say the proportion among all the girls who knew someone who used drugs) and π_2 is the proportion in a second population (say the proportion among all boys).

If the proportions in the two populations are equal then $\pi_1 = \pi_2$ and so $\pi_1 - \pi_2 = 0$. Following the same reasoning as described in Chapter 2, if the confidence interval for the difference in proportions includes 0 then we can be 90, 95 or 99% sure that there is no statistically significant difference in the two population proportions. If the interval does not include 0, then we can conclude with the chosen level of confidence that there is a significant difference between the two population proportions. The approach discussed here to examining the difference between two proportions makes use of the Normal approximation to the binomial. This is justified if the following two conditions are satisfied:

- the total sample size is greater than 20;
- n_1p_1, n_2p_2, $n_1(1 - p_1)$, and $n_2(1 - p_2)$ are all greater than 5.

If either of these conditions is not met then the alternative hypothesis testing approach to examining the difference in two population proportions, described in Chapter 5, should be used. As with estimation with two means we have to consider the same two distinct situations described in Chapter 2, that with independent samples and that with matched samples.

Now do you see what I mean by a significant difference in proportions?

Independent Samples

The algebraic expressions in this section may be too much for you first time round. My advice is the same as before: skim read these pages and come back to them later. The CIA program will anyway do all the hard work for you. The confidence interval for the difference between two independent proportions is:

$$(p_1 - p_2) \pm z \times \text{s.e.}(p_1 - p_2)$$

where p_1 and p_2 are the sample proportions in the two groups and s.e.$(p_1 - p_2)$ is the standard error of the sampling distribution of the difference in two proportions. As usual, z will equal 1.64, 1.96 or 2.58 depending on the required confidence level. The standard error of the difference in sample proportions is:

$$\text{s.e.}(p_1 - p_2) = \sqrt{\frac{p_1(1-p_1)}{n_1} + \frac{p_2(1-p_2)}{n_2}}.$$

I didn't go through a worked example for the difference between two means in Chapter 2 because most statistical packages will do this calculation. However, this is not the case for the difference in population proportions and it's probably a good idea therefore to work through one. We can do this using a real example.

An Example from Practice

In a randomised clinical trial to study the care of asthma sufferers, patients were allocated at random either to an "integrated" care group (GP *and* specialist chest physicians share in the care of the patient) or to a "conventional" care group (patients are cared for by their GP in the usual way). The two groups are thus independent. The percentage of patients who were "very satisfied with their medical care over the past year" was measured for both groups. There were 263 out of 333 such patients in the integrated care group and 284 out of 333 in the conventional care group. The sample proportions were thus: $p_1 = 263/333 = 0.7900$ and $p_2 = 284/333 = 0.8528$. Therefore:

$$\text{s.e.}(p_1 - p_2) = \sqrt{\frac{0.7900(1-0.7900)}{333} + \frac{0.8528(1-0.8528)}{333}} = 0.0296.$$

The point estimate of the difference in proportions is $0.7900 - 0.8528 = -0.0628$ or -6.28%, indicating that more of those assigned to conventional care were very satisfied with their care than those assigned to integrated care. The 95% confidence interval for $(\pi_1 - \pi_2)$, the difference in these proportions is:

$$(0.7900 - 0.8528) \pm 1.96 \times 0.0296$$

$$\text{or} \quad -0.0628 \pm 0.0580.$$

The 95% confidence interval is therefore:

$$-0.1208 \text{ to } -0.0048$$

$$\text{or} \quad -12.08\% \text{ to } -0.48\%.$$

The fact that this interval does not include 0 (just!) means that we can be 95% per cent certain that there *is* a significant difference between the size of the proportions in the two groups who had been "very satisfied" with their care.

Matched Samples

When individuals in the two samples are matched, as they might be, for example, in a case–control study, or in a before–after study, the standard error of the difference in the two sample proportions is different from the independent case. To illustrate how, suppose each person in a group of 12 patients who are waiting for a vasectomy procedure is asked to read a "reassuring" explanatory leaflet*. After they have read the leaflet they are each asked if they feel anxious about what is to happen, and the number who say they do is counted. A similar group of 12 patients, matched in pairs on age, occupational status, number of children, and marital/cohabiting status, is asked to read a leaflet of similar length on the history of the hospital (to act as a control). They are each also asked whether or not they feel anxious. The responses to the questions for each pair of patients is shown in Table 3.2.

Table 3.2: Responses to the question "Do you feel anxious about the vasectomy procedure?" by matched pairs of patients

PAIR	"FEEL ANXIOUS?"	
	Reassuring leaflet	*History leaflet*
1	Y	N
2	N	Y
3	N	N
4	N	N
5	Y	N
6	Y	Y
7	N	Y
8	Y	N
9	N	N
10	Y	N
11	Y	Y
12	Y	N

Table 3.3: Response pairs from Table 3.2

REASSURING LEAFLET	HISTORY LEAFLET	NUMBER IN EACH COMBINATION
Y	Y	2 (*a*)
N	Y	2 (*b*)
Y	N	5 (*c*)
N	N	3 (*d*)

* I know this example fails the conditions on sample size set out earlier for using the Normal approximation to the binomial, but this example is only for the purpose of showing how the formula is applied.

Table 3.4: The information in Table 3.3 cast as a 2 × 2 table

		REASSURING LEAFLET GROUP	
		Y	N
HISTORY LEAFLET GROUP	Y	2 (*a*)	2 (*b*)
	N	5 (*c*)	3 (*d*)

In Table 3.3 the responses are collected into Y-Y, N-Y, Y-N and N-N pairs, which are labelled as *a, b, c* and *d*, and in Table 3.4 they are put into the form of a 2 × 2 table. As you can see, seven of the 12 in the reassuring leaflet group said they were anxious, compared with only four in the history leaflet group. So the sample proportions are:

$$p_1 = 7/12 = 0.5833 \text{ and } p_2 = 4/12 = 0.3333.$$

So the point estimate of the difference in sample proportions, $(p_1 - p_2)$, is 0.5833 – 0.3333 = 0.2500 or 25%. The standard error of the difference in matched population proportions is given by the expression:

$$\text{s.e.}(p_1 - p_2) = \frac{1}{n}\sqrt{b + c - \frac{(b-c)^2}{n}}$$

where n is the number of *pairs*. Using the values from Table 3.4 of $a = 2$, $b = 2$, $c = 5$ and $d = 3$, gives:

$$\text{s.e.}(p_1 - p_2) = \frac{1}{12}\sqrt{2 + 5 - \frac{(2-5)^2}{12}} = 0.2083.$$

The confidence interval estimate of the difference in population proportions is:

$$(p_1 - p_2) \pm z \times \text{s.e.}(p_1 - p_2)$$

or $0.2500 \pm 1.96 \times 0.2083$

or 0.25 ± 0.4083

which is:

-0.1583 to 0.6583

or -15.83% to 65.83%.

Since this interval includes the value 0, we can be 95% sure that there is no statistically significant difference in the proportion of patients in the two groups who felt anxious. The enormous width of this interval is due to the small sample size. As we noted earlier, with samples as small as this we would be better advised to use one of the approaches described in Chapter 5.

Using a Computer to Get Confidence Intervals with Two Proportions

I think that only the CIA package among the four computer packages considered in this book will calculate a confidence interval for the difference between two population proportions, for both the independent (or unpaired) and the matched (or paired) cases. For the unpaired case the commands in CIA are:

Chapter 4 – Proportions and their Differences
2 : Two samples – Unpaired Case

CIA asks you to enter for each sample in turn, (i) the sample size, and (ii) the number of individuals with the feature of interest. The CIA output for the difference in two independent proportions for the conventional versus integrated asthma care study discussed above is shown in Figure 3.2.

```
FIRST SAMPLE:
SAMPLE SIZE: 333
NUMBER WITH FEATURE: 263

SECOND SAMPLE:
SAMPLE SIZE: 333
NUMBER WITH FEATURE: 284

FIRST SAMPLE PROPORTION: 0.790
SECOND SAMPLE PROPORTION: 0.853
OBSERVED DIFFERENCE BETWEEN PROPORTIONS: 0.0631

% CONFIDENCE REQUIRED: 95

--------------------------------------------------

Standard Error of Difference = 0.0296   NORMAL Value = 1.96

95% CONFIDENCE INTERVAL FOR THE DIFFERENCE BETWEEN PROPORTIONS IS:
-0.121 TO -0.00507

--------------------------------------------------

Another level of confidence (Y/N)?
```

Figure 3.2: Output of CIA for 95% confidence interval for the difference in independent proportions of patients responding "very satisfied" in integrated asthma care study

For the matched case, the CIA command sequence is:

Chapter 4 – Proportions and their Differences
3 : Two samples – Paired Case

CIA asks for the number of pairs with (i) the feature present in both samples, (ii) the feature present only in the first sample, (iii) the feature present only in the second sample, and (iv) the feature absent in both samples. The output from CIA for the vasectomy example above is shown in Figure 3.3.

SUMMARY

This chapter has discussed confidence intervals for a population proportion and for the difference between two proportions, for both the independent and matched cases. As far as I am aware the CIA package is the only one offering

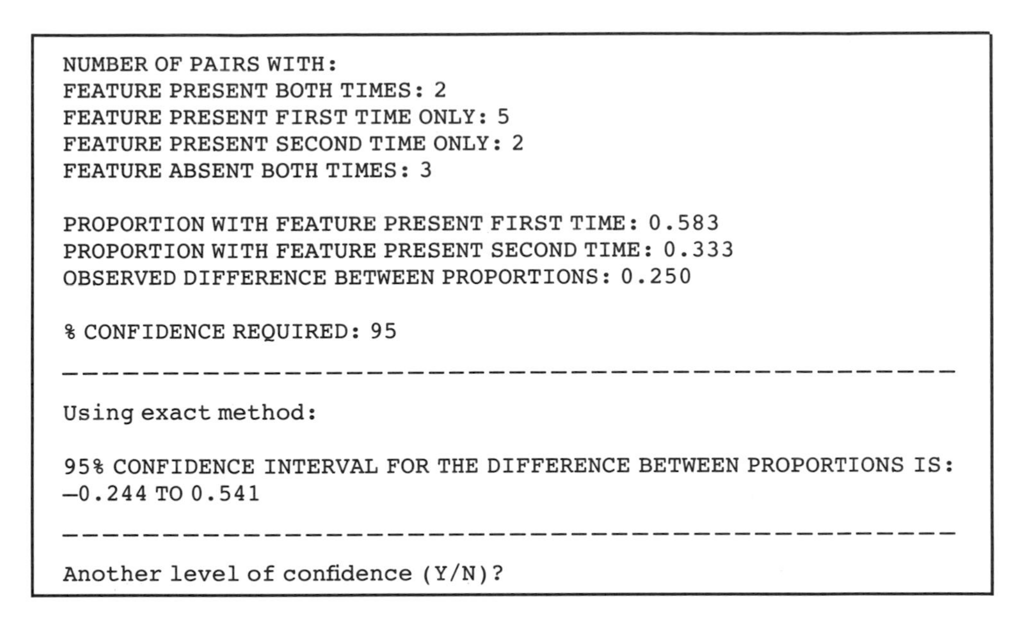

```
NUMBER OF PAIRS WITH:
FEATURE PRESENT BOTH TIMES: 2
FEATURE PRESENT FIRST TIME ONLY: 5
FEATURE PRESENT SECOND TIME ONLY: 2
FEATURE ABSENT BOTH TIMES: 3

PROPORTION WITH FEATURE PRESENT FIRST TIME: 0.583
PROPORTION WITH FEATURE PRESENT SECOND TIME: 0.333
OBSERVED DIFFERENCE BETWEEN PROPORTIONS: 0.250

% CONFIDENCE REQUIRED: 95

-------------------------------------------------

Using exact method:

95% CONFIDENCE INTERVAL FOR THE DIFFERENCE BETWEEN PROPORTIONS IS:
-0.244 TO 0.541

-------------------------------------------------

Another level of confidence (Y/N)?
```

Figure 3.3: Output of CIA for 95% confidence interval for the difference in matched proportions of patients expressing anxiety in vasectomy example

direct computation of confidence intervals for the one-sample and two-sample cases.

EXERCISES

3.1 In an investigation of reduced bone density in women taking depot medroxy-progesterone (DMPA) as a contraceptive, a sample of 14 women were recruited who had taken DMPA but had stopped. Four of these women smoked. Use an appropriate procedure to calculate a 95% confidence interval estimate of the proportion of women in this population who smoked. Comment on your result.

3.2 Use the data[4] in Table 3.5 which shows the number of successful suicides in Copenhagen in 1980 to calculate 95% confidence interval estimates for the population proportion of male and of female suicides who were schizophrenic. Compare the male and female confidence intervals.

3.3 In a study into the knowledge and experience of young people with illicit drugs[2], 29 out of a sample of 228 school children aged 14 to 15 reported knowing someone who used the drug "ecstasy". Use an appropriate procedure to calculate 95 and 99% confidence interval estimates of the corresponding population proportion. Comment on your results.

3.4 In a clinical trial to assess the effectiveness of health checks in reducing risk factors for cardiovascular disease[5], the treatment group alone received counselling and advice on changing to healthier lifestyles. One year on the resulting percentage difference of the treatment group from the control group, and 95% confidence intervals for these percentage differences, for a number of

Table 3.5: Number of successful suicides by sex and diagnosis in Copenhagen in 1980

DIAGNOSIS	MALE	FEMALE
Schizophrenia	2	4
Affective psychosis	1	7
Psychogenic psychosis	0	2
Neurosis	2	7
Personality disorder	12	8
Alcoholism	16	12
Drug addiction	2	1
Other	2	7
Not mentally ill	13	14

Table 3.6: Percentage difference from control and 95% confidence intervals for the differences between treatment and control groups (females only) across a number of factors in a study into the effectiveness of health checks in reducing cardiovascular disease

FACTOR	PERCENTAGE DIFFERENCE FROM CONTROL AND 95% CONFIDENCE INTERVAL OF DIFFERENCE
% smokers	1.7 (–1.3 to 4.7)
% stopped smoking in previous year	2.2 (–0.4 to 4.9)
% with diastolic b.p. ≥ 100 mmHg	1.1 (0.1 to 2.1)
% with total cholesterol ≥ 8 mmol/l	3.9 (2.2 to 5.7)
% with body mass index ≥ 30	1.6 (–0.9 to 4.0)

factors in female subjects, are shown in Table 3.6. Comment on what these confidence intervals show.

3.5 Calculate the 99% confidence interval for the difference in population proportions for the clinical trial on the integrated care of asthma patients described above on page 45. Compare with the 95% value calculated there.

3.6 In an international multicentre frequency-matched case–control study[6] of the effects of dietary isomeric *trans* fatty acids on coronary heart disease, the subjects were men aged 70 years or younger from 10 study centres examined during 1991 and 1992. Cases were 742 men with acute myocardial infarction (AMI); controls were 757 men without a history of AMI recruited from the population in the catchment area and frequency matched for age. Researchers calculated the differences between cases and controls in the percentage of subjects with two major risk factors (as possible confounders); percentage smoking and percentage with low socio-economic status. The 95% confidence intervals for the percentage differences in the first two of these for three out of the 10 centres are shown in Table 3.7. Comment on what these values show.

Table 3.7: Differences between cases and controls in the percentage with two major risk factors (as possible confounders) together with the 95% confidence intervals for the differences in these percentages, in a study into the effect of isometric *trans* fatty acids on coronary heart disease

CENTRE	PERCENTAGE SMOKERS			PERCENTAGE LOW SOCIO-ECONOMIC STATUS		
	Cases	*Controls*	*95% CI for difference*	*Cases*	*Controls*	*95% CI for difference*
Helsinki	63	25	23 to 53	47	29	1 to 35
Jerusalem	52	33	1 to 37	8	12	–15 to 7
Zeist	55	38	0 to 34	49	45	–14 to 22

REFERENCES

1. *Geigy Scientific Tables*, 1982.
2. Wright, J. D. and Pearl, L. (1995) Knowledge and experience of young people regarding drug misuse, 1969–94. *BMJ*, **310**, 20–4.
3. GRASSIC, Grampian Asthma Study of Integrated Care (1994) Integrated care for asthma: a clinical, social, and economic evaluation. *BMJ*, **308**, 550–64.
4. Nordentoft, M. *et al.* (1993) High mortality by natural and unnatural causes: a ten-year follow-up study of patients admitted to a poisoning treatment centre after suicide attempts. *BMJ*, **306**, 1637–41.
5. Imperial Cancer Research Fund OXCHECK Study Group (1994) Effectiveness of health checks conducted by nurses in primary care: results of the OXCHECK study after one year. *BMJ*, **308**, 308–12.
6. Aro, A. *et al.* (1995) Adipose tissue isomeric *trans* fatty acids and risk of myocardial infarction in nine countries: the EURAMIC study. *The Lancet*, **345**, 273–76.

4

ESTIMATING THE POPULATION MEDIAN

❑ The sample median as the best point estimator of the population median ❑ Use with ordinal data and skewed metric data ❑ Confidence intervals for the median; with small samples; with large samples ❑ Confidence intervals for any percentile ❑ Estimating the difference between two population medians ❑

THE MEDIAN IS THE MESSAGE

There are three circumstances when we might choose the median as our preferred measure of location:

- if the data are ordinal;
- if the data are metric but the sample size is very small;
- if the data are metric but are markedly skewed.

Estimation of the population median is a non-parametric procedure since it does not depend on the variable in the population in question having a Normal distribution. As we might have guessed, the sample median m is an unbiased point estimator of the population median M*. However, as with estimates of the population mean and population proportion, if we want to attach some degree of confidence to the point estimate, we have to calculate the appropriate confidence intervals. There are two separate cases to consider: first when we are dealing with a small sample, and second when the sample is large.

Small Samples, *n* up to 50

The small sample procedure, for sample sizes up to 50, makes use of values found using rather complex calculations based on the binomial distribution. However, these can be avoided because the necessary values have already been calculated and are given in Table A4 in the Appendix. A small section of Table A4 is shown in Table 4.1. Using this table to obtain confidence intervals for a median is very simple as we shall now see.

* There are no commonly accepted symbols for the sample and population median. I have chosen to use m and M respectively in this book.

An Example from Practice

The data in Table 4.2 give the ages of subjects in a study[1] into the effects of nitrogen dioxide on the airway response to inhaled allergens in 10 asthmatic patients. We want an estimate of the "average" age of the population of asthmatics of which this sample is representative. Since age is a metric variable we would normally choose the mean as our measure of average. There are two reasons why this is not appropriate with these data:

Table 4.1: Small section of Table A4 showing upper and lower limits for interval estimates of the population median for small samples (up to $n = 50$)

	TARGET CONFIDENCE LEVELS								
	90%			95%			99%		
n	L_l	L_u	*Actual* (%)	L_l	L_u	*Actual* (%)	L_l	L_u	*Actual* (%)
9	3	7	82.03	2	8	96.09	1	9	99.61
10	3	8	89.06	**2**	**9**	**97.85**	1	10	99.80
11	3	9	93.46	2	10	98.83	1	11	99.90

Table 4.2: Age of 10 subjects in an investigation of the effect of nitrogen dioxide on airway responses to inhaled allergens in asthma sufferers

SUBJECT	AGE (years)	AGE (ascending order)
1	29	19
2	47	23
3	28	24
4	19	24
5	24	24
6	27	27
7	34	27
8	23	28
9	24	29
10	24	47

Not only is this a small sample ($n = 10$), but the data appear to be skewed (note the presence of the outlier at 47 years). The sample median is therefore preferred as the most appropriate measure of location*. Using the procedure described in *Statistics from Scratch*, the sample median age is the average of the 5th and 6th observations, i.e. half-way between 24 and 27, which is 25.5 years (incidentally the sample *mean* age is 27.9 years, the presence of the outlier of 47 years drags the mean upwards).

The lower and upper limits for the 95% confidence interval of the population median age are found from Table 4.1 on the $n = 10$ row. Here the values of $L_l = 2$ and $L_u = 9$ are found in the column corresponding to a confidence level of 95%. In other words, the lower limit is the value of the 2nd observation, i.e. 23 years, and the upper limit the value of the 9th observation, i.e. 29 years. So the 95% confidence estimate of population mean

* Recall from *Statistics from Scratch* that the median is equal to the value of the middle observation, i.e. of the ½(n+1)th observation. In this case, ½(n+1) = ½(11+1) = 5.5.

age is (23, 29) years. Notice from Table 4.1 that the confidence level actually achieved is 97.85%, even though we wanted a confidence level of 95%. The discrepancy between the actual confidence level and the *target* confidence level is caused by the fact that the values L and U are discrete, whereas age is, of course, a continuous variable.

Large Samples, *n* over 50

For samples where n is greater than 50, Table A4 cannot be used but good approximations can be found using the Normal distribution. The procedure is as follows:

- *Step 1:* Choose the target level of confidence: 90%, 95% or 99%.
- *Step 2:* Arrange the sample observations in ascending order.
- *Step 3:* Find the value of z corresponding to the target confidence level, i.e. either 1.65 (for 90%), 1.96 (for 95%), or 2.58 (for 99%).
- *Step 4:* Calculate the position of the lower bound of the interval estimate, L, using the expression:

$$L = \frac{1}{2}\,n - \frac{1}{2}\,z\sqrt{n}.$$

- *Step 5:* Round L to the nearest whole number.
- *Step 6:* Then the position of the upper bound of the interval estimate, U, is found from:

$$U = 1 + n - L.$$

- *Step 7:* Determine the value of the sample observation in the Lth and Uth positions in the ascending data. These are the lower and upper bounds respectively of the interval estimate.
- *Step 8:* Find the actual confidence level as follows. Calculate the value of:

$$z = \frac{2L - n - 1}{\sqrt{n}}.$$

  Use the z table (Table A1 in the Appendix) to find the area in the tail of the z distribution beyond this value of z. Call this area $\alpha/2$. The actual confidence level is then $= 100(1 - 2 \times \alpha/2)\%$.

As an example suppose we require a 99% confidence interval estimate for the median score of 60 Injury Severity Scores (ISS) made on a sample of 60 patients admitted to a hospital's A&E department. The data, arranged in ascending order, are shown in Table 4.3. (Note that ISS scores are ordinal.)

The best point estimator of the population median ISS score is the sample median ISS score. This is the score in the $\frac{1}{2}(n + 1)$th position, i.e. in the 30.5th position. Thus the sample median ISS score equals 12.5. Following the above procedure:

- *Step 1:* The confidence level is 99%.
- *Step 2:* Data are already in ascending order.
- *Step 3:* For 99%, $z = 2.58$.
- *Step 4:* The position of the lower bound of the interval estimate is:

$$L = \frac{60}{2} - \frac{2.58 \times \sqrt{60}}{2} = 20.008.$$

Table 4.3: Injury Severity Scores (ISS), arranged in ascending order, for a sample of 60 patients admitted to a hospital's A&E department

ISS SCORE	RANK	ISS SCORE	RANK	ISS SCORE	RANK
1	1	10	21	16	41
1	2	10	22	16	42
2	3	10	23	17	43
2	4	10	24	17	44
2	5	12	25	17	45
3	6	12	26	18	46
3	7	12	27	19	47
4	8	12	28	20	48
4	9	12	29	20	49
5	10	12	30	20	50
6	11	13	31	24	51
8	12	13	32	28	52
8	13	13	33	32	53
8	14	13	34	38	54
8	15	14	35	45	55
9	16	14	36	46	56
9	17	14	37	55	57
9	18	15	38	66	58
10	19	15	39	70	59
10	20	15	40	70	60

- *Step 5:* Rounding to nearest integer gives the position of the lower bound as $L = 20$
- *Step 6:* The position of the upper bound of the interval is:

$$U = 1 + 60 - 20 = 41.$$

- *Step 7:* The value of the ISS score in the 20th position is 10; the ISS score in the 41st position is 16. The 99% confidence interval for the population median ISS score is therefore from 10 to 16.
- *Step 8:* To find the actual confidence level calculate:

$$z = [(2 \times 20) - 60 - 1]/\sqrt{60} = -2.711.$$

Find the area in the tail of the standard Normal distribution to the left of $z = -2.711$. From the z table the area between $z = 0$ and $z = -2.711$ is 0.4966. The area in the tail is therefore 0.5 – 0.4966 = 0.0034*. So $\alpha/2 = 0.0034$. The actual confidence level is $100(1 - 2 \times 0.0034)\% = 99.32\%$.

An Example from Practice

Clinical auditors measured the time (in days) spent in hospital by every one of the 143 patients admitted during a six-week period ($n = 143$). The sample data appeared to be highly positively skewed with a minimum stay of 0 days (patients discharged on the same day as they were admitted) to a maximum of 186 days, with more than half of the patients staying less than 20 days. Accordingly, it was decided to use the median as the most

* Remember that the total area under the z curve is 1.0 so the area in each half is 0.5.

representative measure of average length of stay. The median stay was found to be 15 days, with a targeted 95% confidence interval of population median stay of (11 days to 19 days). The actual level of confidence achieved was 95.4%.

As a matter of interest, the mean stay was found to be 26.53 days with a 95% confidence interval estimate of (21.041 days to 32.019 days). The presence of the long right-hand tail in the distribution noticeably affects the value of the mean, and the precision of the interval estimate which is considerably wider than that for the median (eight days compared with 11 days).

In general, if the population is Normal, the mean will be equal to the median (because of course the distribution is symmetric) and the intervals are both estimating the middle of the distribution. In these circumstances the interval for the mean will be narrower than that for the median since the latter interval is valid for all populations regardless of the shape of the distribution. When the distribution is not symmetric, the confidence interval for the mean may be wider than that for the median.

Interval Estimates for Any Percentile

The confidence interval for any percentile q can be calculated using a similar method to that described for the median above. The lower bound of the interval for the qth percentile is the sample observation in the position of the integer nearest to the value of L where:

$$L = nq - z \times \sqrt{nq(1-q)}$$

and the upper bound is the value of the sample observation in the position of the integer nearest to U where:

$$U = 1 + nq - z \times \sqrt{nq(1-q)}.$$

In other words, to find a 95% confidence interval estimate for the 75th percentile, $q = 0.75$ and $z = 1.96$ would be substituted in the above equations. An example is left to the exercises at the end of this chapter.

Using a Computer to Get Interval Estimates of the Population Median

Minitab can be used to calculate interval estimates of the median using the **WInterval** command, the output from which is shown in Figure 4.1 for the median age of male and female suicides (raw data in Table 2.4). If the data are entered into columns 1 and 2, say, of the Minitab worksheet and named "males" and "females", the necessary commands are:

Stat
  Nonparametrics
    1-Sample Wilcoxon
      Select c1 c2
        ⊙ Confidence interval
          (Accept of change the default 95%)
            OK

```
MTB > WInterval 95.0 ''males'' ''females''

Wilcoxon Signed Rank Confidence Interval

                    Estimated    Achieved
             N       median     confidence   Confidence interval
Males       99       38.50        95.0          (35.0, 42.5)
Females     20       49.00        95.0          (40.0, 58.5)
```

Figure 4.1: Output from the Minitab **WInterval** command for 95% confidence intervals for the median age of male and female suicides

SPSS does not produce interval estimates for a median directly but the CIA program does. The CIA command sequence is:

Chapter 8 – Some Non-parametric Analyses
1 : Single Sample – Median

The CIA program asks you if you want interval estimates based on the binomial distribution or the Wicoxon procedure. It is best to choose the binomial approach. If you choose the Wilcoxon method you will be confronted by a screen which at first sight seems unintelligible*.

The program identifies the sample observations whose values are the lower and upper bounds of the confidence interval estimate, when the sample values in Table 2.4 are sorted† into ascending order (Figure 4.4). These are the 40th and 60th for men, and the 6th and 15th for women (shown bold). This gives 95% confidence interval estimates of the median age for male suicides of (32, 41) years, and for female suicides of (38, 63) years. The CIA output is not shown.

ESTIMATING THE DIFFERENCE BETWEEN TWO MEDIANS

If we wish to compare the locations of two distributions but the sample sizes are small and we cannot be sure that the differences are Normal (i.e. the *parametric* assumptions required for comparing two means using the method based on the t distribution described in Chapter 2 are not satisfied) we may wish to play safe and compare the two medians instead. The method described below for calculating a confidence interval for the difference in two population medians does not require the assumption of Normal distributions and hence is a *non-parametric* procedure. The price paid is that the methods produce slightly less precise results, i.e. slightly wider confidence intervals.

The big problem with finding a confidence interval for the *difference* between two population medians is that if the data are ordinal we have seen that it makes little sense to apply any of the basic rules of arithmetic, and that includes subtracting (i.e. differencing) the values in one sample from those in the other sample. This we obviously would have to do if we were trying to measure the difference between them. For this reason we must approach calculations of the difference

* In fact the output requires you to refer to the *Geigy Scientific Tables*, which you may not have to hand.
† In Minitab the sorted values are obtained using commands **Manip**, **Sort**, **Select c1 c2**, **Store sorted columns in c3 c4**.

Table 4.4: Age of suicides (raw data in Table 2.4) sorted into ascending order

ROW	MALES (age)	FEMALES (age)	ROW	MALES (age)	ROW	MALES (age)
1	17	17	34	30	67	45
2	18	21	35	30	68	46
3	20	25	36	30	69	47
4	20	26	37	31	70	47
5	21	32	38	31	71	48
6	21	**38**	39	31	72	48
7	21	39	40	**32**	73	49
8	21	42	41	32	74	50
9	22	43	42	32	75	50
10	22	49	43	33	76	51
11	22	52	44	33	77	51
12	22	54	45	34	78	52
13	23	55	46	34	79	53
14	23	59	47	34	80	54
15	24	**63**	48	35	81	54
16	24	68	49	36	82	57
17	24	72	50	36	83	57
18	24	73	51	36	84	60
19	25	73	52	36	85	60
20	25	79	53	36	86	61
21	26		54	37	87	63
22	26		55	38	88	64
23	26		56	38	89	65
24	27		57	39	90	65
25	27		58	40	91	66
26	27		59	40	92	68
27	27		60	**41**	93	68
28	28		61	41	94	68
29	28		62	43	95	69
30	30		63	44	96	72
31	30		64	45	97	78
32	30		65	45	98	79
33	30		66	45	99	86

between medians when the data are ordinal with great caution. I will not be considering this possibility in this book. If we are dealing with metric data the problem doesn't arise since metric values are truly numeric and we can subtract them to find differences between samples.

The method I am going to use below requires us to assume that the data in the two samples are metric, and that the two distributions are identical in shape (and thus have the same standard deviations), but that they may have different locations (which is what we want to find out)*. I am not going to describe the method which would be tedious because of its length and complexity. However, we will examine its use in two packages, CIA and Minitab (SPSS does not appear to include this facility), both of which use an approach based on what is known as the *Mann–Whitney* procedure. This approach is more often used in the context of a

* The fact that the two distributions are identical in shape means that this procedure examines the difference between the two population means as well as the two medians.

hypothesis test and will be described in greater detail in Chapter 6. Once again we have to distinguish between two matched or paired samples and two independent samples.

Independent Samples

As an example of the *independent* samples case, consider Table 4.5 which contains data on the age of male and female students at Oxford University committing suicide between 1976 and 1990.[2] Although the data is metric, the samples are small and we cannot therefore say anything about their shape. Accordingly, a non-parametric procedure seems most appropriate.

Table 4.5: Ages of male and female students at Oxford University committing suicide between 1976 and 1990

MALES	18	19	20	20	21	21	21	21	21	22	22	22	22	23	23	25	(n = 16)
FEMALES	21	22	23	23	24	(n = 5)											

Using CIA with Independent Samples

To use the CIA program we follow the command sequence:

Chapter 8 – Some Non-parametric Analyses
2: Two Samples – Unpaired Case – Differences between Medians

The output using the data in Table 4.5 is shown in Figure 4.2 for two confidence levels. Notice that, as is invariably the case with confidence intervals for medians, we don't get exactly the confidence level we require because although age is a continuous variable the median confidence interval procedure assumes discrete data, and the confidence level is thus approximated. In this example we actually get a 96% level instead of the desired 95% and a 99.2% level instead of 99%.

The point estimate of the difference between the median ages of males and females is 1 year, but since both confidence intervals include zero, (–3.00, 0) and (–4.00, 1.00), we have to conclude that there is no statistically significant difference in the median ages of the population of male and female student suicides as represented by these two samples.

Using Minitab with Independent Samples

Minitab produces a confidence interval for the median as well as the results of the Mann–Whitney hypothesis test (which we can ignore for the moment). Assuming we have the male and female ages in columns 1 and 2, the command sequence is:

Stat
Nonparametrics
Mann-Whitney
Select c1
Select c2
(Accept default 95%)
OK

```
FIRST SAMPLE:
SAMPLE SIZE: 16

SECOND SAMPLE:
SAMPLE SIZE: 5

DIFFERENCE BETWEEN MEDIANS = -1.00

% CONFIDENCE REQUIRED: 95

----------------------------------------------

K = 16

96% CONFIDENCE INTERVAL FOR THE DIFFERENCE BETWEEN MEDIANS IS:
from -3.00 to 0

----------------------------------------------

Another level of confidence (Y/N)?

FIRST SAMPLE:
SAMPLE SIZE: 16

SECOND SAMPLE:
SAMPLE SIZE: 5

DIFFERENCE BETWEEN MEDIANS = -1.00

% CONFIDENCE REQUIRED: 99

----------------------------------------------

K = 10

99.2% CONFIDENCE INTERVAL FOR THE DIFFERENCE BETWEEN MEDIANS IS:
from -3.00 to 1.00

----------------------------------------------

Another level of confidence (Y/N)?
```

Figure 4.2: Output of CIA for 95% and 99% confidence intervals for the difference in median ages of male and female student suicides

If the 95% default confidence level is acceptable click on **OK** (otherwise specify a different confidence level). The output for both 95 and 99% confidence levels is shown in Figure 4.3 (the accompanying hypothesis test results have been omitted from this output). The results are the same as for the CIA program and confirm no significant difference in median ages.

Paired or Matched Samples

We can use CIA and Minitab to calculate confidence intervals for the difference in the medians of paired or matched samples. Both packages use a method based on the *Wilcoxon signed rank* procedure (again used mostly in the context of hypothesis testing and to be discussed in more detail in Chapter 6).

```
MTB > Mann–Whitney 95.0 ''males'' ''females'';
SUBC> Alternative 0.

Mann–Whitney Confidence Interval and Test

Males      N = 16    Median = 21.000
Females    N =  5    Median = 23.000

Point estimate for ETA1-ETA2 is     –1.000

95.7 pct c.i. for ETA1-ETA2 is (–3.000, –0.000)


MTB > Mann–Whitney 99 ''males'' ''females'';
SUBC> Alternative 0.

Mann–Whitney Confidence Interval and Test

Males      N = 16    Median = 21.000
Females    N =  5    Median = 23.000

Point estimate for ETA1-ETA2 is     –1.000

99.1 pct c.i. for ETA1-ETA2 is (–3.999, 1.000)
```

Figure 4.3: Using Minitab's Mann–Whitney procedure to calculate 95% and 99% confidence intervals for the difference in the ages at which male and female Oxford University students committed suicide from 1976 to 1990 (the test results have been omitted from the output)

Using Minitab with Paired or Matched Samples

As an example of the use of Minitab in the *paired* case, consider once again the data in Table 2.7 recording the duration of the first and fourth seizures experienced by 40 patients undergoing ECT. The Wilcoxon procedure is applied to the *differences* between each pair of sample values. In Minitab we can use the **let** command (as we did with matched means in Chapter 2) to calculate a column of difference values. For example, if the values for the duration of the first and fourth seizures are in columns c1 and c2 respectively, we would simply type in the sessions window:

let c3 = c1 – c2
name c3 'diff'

which would place the difference values in column c3 and name the column *diff*.

To get the confidence interval for the difference in the median duration of the first and fourth ECT seizures, the commands in Minitab are:

Stat
  Nonparametrics
    Select diff
      ⊙ Confidence interval
        (Accept 95% default)
          OK

```
MTB > WInterval 95.0 ''diff''.

Wilcoxon Signed Rank Confidence Interval

                  Estimated    Achieved
          N       median       confidence   Confidence interval
diff      40      4.50         95.0         (0.00, 9.00)
```

Figure 4.4: Using Minitab's Wilcoxon signed rank procedure to calculate a 95% confidence interval for the difference between the median duration of first and fourth ECT seizures (raw data in Table 2.7)

The output is shown in Figure 4.4. The point estimate of the difference between the median duration of the first and fourth seizures is 4.5 seconds, but the confidence interval (0.00 to 9.00 seconds) includes zero, which implies that there is in fact no significant difference in the median duration of the first and fourth ECT seizures.

An Example from Practice

In a study into the effectiveness of a self-help manual with patients suffering from bulimia nervosa,[3] subjects in a study were randomly assigned to one of three groups: a group given the self-treatment manual who were encouraged to work through the book carefully, doing the exercises; a group receiving eight sessions of cognitive behavioural therapy; and a waiting list group who were told they would have to wait eight weeks for their therapy to begin. Bulimic symptoms were measured before and after eight weeks, and outcome measures included bingeing; vomiting; using laxatives, diuretics, or diet pills; fasting; abnormal dietary patterns; and so on.

This study reported the difference in the median symptom scores before and after the eight-week period. Note that these symptom scores are *ordinal*. Some of the results are shown in Table 4.6.

Table 4.6: Difference in the median bulimic symptom scores before and after the eight-week period in a controlled trial of self-help manual (25th and 75th percentile scores in brackets)

	MANUAL GROUP			COGNITIVE BEHAVIOURAL THERAPY GROUP		
	Before treatment	*After treatment*	*Median change (95% conf. int.)*	*Before treatment*	*After treatment*	*Median change (95% conf. int.)*
Bingeing	3(1,4)	1(0,3)	1(0 to 2)	4(2,4)	1(0,2.5)	2(0 to 3)
Vomiting	3(0,5)	1(0,3)	1(0 to 1)	3(0,4)	0(0,2.5)	1(0 to 2)
	WAITING LIST GROUP					
	Before treatment	*After treatment*	*Median change (95% conf. int.)*			
Bingeing	3(2,4.75)	3(1,4)	0(−1 to 1)			
Vomiting	1(0,4.75)	1(0,3)	0(0 to 1)			

Although the point estimates of the difference in median symptom scores at the beginning and end of the eight-week period showed decreases in symptoms for both the manual and the cognitive behavioural therapy groups (no difference in the waiting list group), the confidence intervals for these differences are indicative of no significant improvement in either group (since they all include zero).

However, in view of the fact that this method involves the calculation of differences in *ordinal* data (recall that ordinal data do not permit arithmetic operations, such as subtraction), these results should be treated with caution.

SUMMARY

This chapter has discussed confidence intervals for the population median and for the difference in two population medians for both the matched and independent samples cases. Methods available for calculations in this context are limited by the fact that ordinal data do not allow for arithmetic manipulation. This means, for example, that we did not consider confidence intervals for the difference in medians for other than samples with metric data.

EXERCISES

4.1 Calculate the 95% confidence interval estimates for the population (a) median, and (b) 25th and 75th percentile, Waterlow Pressure Sore Risk Assessment scores using the 40 sample Waterlow scores shown in Table 4.7. Comment on your results.

4.2 Find the 95% confidence interval estimate for the median of the 60 ISS scores shown in Table 4.3. Comment on your result.

4.3 Calculate the 95 and 99% confidence interval estimates for the population median number of co-workers of subjects in a study into passive smoking and heart disease, using the data in Table 4.8. Comment on your result. What is the population referred to here?

4.4 Look at the suicide data in Table 2.4. Why do you think the median might be a more appropriate measure of location for the age of male suicides than the mean. Calculate 95 and 99% confidence interval estimates of the population median age of male suicides. What population are we talking about here?

4.5 Calculate a 95% confidence interval for the difference in the median ages of male and female suicides for the data in Table 2.4.

4.6 The data in Table 4.9 refer to the Barthel score of 29 elderly patients in a District General Hospital before and after treatment. Why are these two samples matched? Calculate 95 and 99% confidence intervals for the difference in median before and after scores. Explain why data such as these are not in fact suitable for such a calculation.

Table 4.7: Waterlow scores from a pressure sore prevalence audit (Key: column 1 = patient number; column 2 = gender; column 3 = age group; column 4 = support surface (mattress type): 01 = ordinary mattress, 02 = Vaperm, 03 = Spenco, 09 = other; column 5 = number of pressure sores; column 6 = Waterlow score)

PATIENT NO.	SEX	AGE GROUP	SUPPORT SURFACE	NO. OF SORES	WATERLOW SCORE
1	M	87–97	02	1	22
2	M	65–75	01	0	13
3	M	65–75	01	0	9
4	F	65–75	02	0	18
5	F	76–86	09	0	19
6	F	76–86	02	1	20
7	F	76–86	02	0	21
8	M	65–75	02	0	14
9	M	65–75	02	0	16
10	M	65–75	02	0	13
11	F	76–86	09	0	15
12	F	65–75	09	0	16
13	F	76–86	09	0	9
14	F	76–86	09	0	15
15	F	87–97	03	4	27
16	F	65–75	02	0	25
17	F	87–97	03	1	23
18	F	65–75	09	4	31
19	F	65–75	01	1	16
20	F	65–75	01	0	9
21	F	87–97	03	2	24
22	F	76–86	01	0	11
23	F	65–75	03	0	11
24	F	65–75	03	1	14
25	F	65–75	01	0	6
26	F	76–86	03	2	14
27	F	65–75	01	3	23
28	F	65–75	01	0	16
29	F	65–75	01	2	17
30	F	65–75	01	0	12
31	F	65–75	01	0	10
32	F	65–75	01	0	14
33	F	76–86	03	0	18
34	F	65–75	01	0	15
35	M	65–75	03	1	17
36	F	65–75	01	0	11
37	F	65–75	03	0	12
38	F	65–75	01	0	6
39	F	65–75	01	0	13
40	M	65–75	01	0	6

Table 4.8: Number of smoking co-workers for a sample of 45 non-smoking females

SUBJECT	1	2	3	4	5	6	7	8	9	10	11	12	13	14	15
SMOKERS	2	0	5	15	1	8	5	0	0	9	10	4	4	2	3
SUBJECT	16	17	18	19	20	21	22	23	24	25	26	27	28	29	30
SMOKERS	5	1	14	7	9	2	3	0	0	5	8	1	6	0	6
SUBJECT	31	32	33	34	35	36	37	38	39	40	41	42	43	44	45
SMOKERS	0	4	7	5	2	2	6	3	1	0	1	1	3	2	0

Table 4.9: Before and after Barthel scores for a group of elderly patients

PATIENT	BEFORE	AFTER	PATIENT	BEFORE	AFTER
1	11	19	16	7	7
2	10	16	17	14	20
3	0	17	18	14	18
4	4	3	19	12	13
5	6	16	20	12	12
6	8	11	21	16	12
7	13	15	22	20	19
8	6	15	23	8	4
9	6	10	24	9	13
10	12	19	25	10	17
11	16	17	26	12	9
12	1	10	27	8	12
13	8	16	28	10	17
14	13	17	29	2	11
15	8	5			

REFERENCES

1. Tunnicliffe, W. S. *et al.* (1994) Effect of domestic concentrations of nitrogen dioxide on airway responses to inhaled allergens in asthmatic patients. *The Lancet*, **344**, 1733–5.
2. Hawton K. *et al.* (1995) Suicide in Oxford University students, *British Journal of Psychiatry*, **166**, 44–6.
3. Treasure, J. *et al.* (1994) First step in managing bulima nervosa: a controlled trial of therapeutic manual. *BMJ*, **308**, 686–9.

5

HYPOTHESIS TESTING: NOMINAL VARIABLES

❑ HYPOTHESIS TESTING AND INTERVAL ESTIMATION ❑ THE NULL AND ALTERNATIVE HYPOTHESIS ❑ TYPES OF ERROR ❑ POWER OF A TEST ❑ THE SIGNIFICANCE LEVEL OF THE TEST AND THE *P*-VALUE; ACCEPTING OR REJECTING THE NULL HYPOTHESIS ❑ ONE- AND TWO-TAILED TESTS ❑ ONE-SAMPLE TESTS: THE BINOMIAL TEST; THE CHI-SQUARED GOODNESS-OF-FIT TEST ❑ TWO-SAMPLE TESTS, MATCHED SAMPLES: MCNEMAR'S TEST; INDEPENDENT SAMPLES: CONTINGENCY TABLES AND THE CHI-SQUARED 2 × 2 TEST; FISHER'S TEST ❑

TO TEST OR NOT TO TEST, THAT IS THE QUESTION

Suppose you work for the blood donation service and decide to site a "Blood Donation" caravan in an out-of-town shopping centre. In an idle moment you wonder about the relative proportions of men and women who come forward to donate blood. Vlad, your colleague, who seems to be a bit of an expert on these matters, thinks that in general the proportions are about the same, but you are not so sure, although you don't really know whether there is a greater proportion of men or of women. You agree to settle the question on the basis of the numbers of men and women coming forward on the first day that the caravan is open for business. This will be your sample, and the number of women in it will allow you to calculate the sample proportion. To make things interesting you have a small bet on the outcome, the loser buys the cakes.

What Vlad is saying is that the population proportion of female donors, π_{female}, is equal to 0.5 (i.e. 50% or a half of all donors). What you're saying is that π_{female} is not 0.5. You could decide the issue in one of two ways. You could use the sample proportion to calculate a confidence interval for π_{female} and see if this interval includes the value 0.5. If it does, Vlad would be proved right and win the bet, if it doesn't you would win. The other possibility is to perform a hypothesis test, using the procedures described below.

In Chapter 2 we saw that if possible the confidence interval approach should always be used in preference to hypothesis testing. A confidence interval, in addition to its main purpose of giving us the 95 or 99% range of values in which we can be confident of finding the population parameter, can *also*, as we have seen, be used to do a hypothesis test. For example, if we are investigating whether there is a significant difference between two means, two medians, or two proportions, we can examine the confidence interval of the difference between them; and if this interval includes 0 then we can conclude that they are not significantly different. The same applies to measures of relative risk and the odds ratio; if

confidence intervals for these include the value 1 then there is no significant change in the risk from exposure to the risk factor.

Despite the arguments in favour of confidence intervals, this chapter and the two which follow are given over to a discussion of the more commonly used hypothesis tests. The reasons for this are as follows:

- because the appropriate software for some confidence intervals is not always as readily available as it is for the equivalent hypothesis tests;
- because there are some tests for which it is not easy to find equivalent or suitable confidence intervals;
- because some hypothesis tests are still widely preferred, e.g. chi-squared, to their equivalent confidence interval of proportions;
- because it is important for reading the literature where hypothesis tests are still widely reported;
- because some individuals are happier working with what they know best.

Even so, since Minitab, SPSS and Excel (and many other packages) all offer easy-to-use hypothesis testing programs, it is unlikely that these will often be performed by hand. I will not therefore describe the procedures in any more detail than is absolutely necessary to understand, first, which test is the most appropriate for any given problem; second, how to use a computer program to perform it; and third, how to interpret the output you get from that program.

THE NULL HYPOTHESIS

Consider again our blood donors example. In the language of hypothesis testing, the assumption that Vlad makes about the relative proportions of men and women donors, i.e. that they each account for half, or 50%, of all donors, would be called the *null hypothesis*. A null hypothesis is usually denoted as H_0 (pronounced "aitch nought"). In this example the null hypothesis would be written:

$$H_0: \pi_{\text{female}} = 0.5$$

where π_{female} is the *population* proportion of female would-be blood donors (note that we could equally have written this null hypothesis as H_0: $\pi_{male} = 0.5$). Of course we have to allow for the possibility that Vlad is wrong and you are right, i.e. that H_0 is not true, and that the proportions of female donors is *not* 0.5. So π_{female} may be more than 0.5, it may be less. These alternative possibilities are represented by what is known as the *alternative hypothesis*, usually denoted as H_1. In this example the alternative hypothesis would be written:

$$H_1: \pi_{female} \neq 0.5$$

where the symbol ≠ means "not equal to". Essentially, hypothesis testing amounts to asking whether or not the sample data offer evidence in support of the null hypothesis. If they do we can *accept* (or are not able to reject) the null hypothesis, if they don't we *reject* the null hypothesis. Thus the crucial question is: what sort of sample data would enable us to accept or cause us to reject H_0?

In most situations, we hope to be able to *reject* the null hypothesis, which will usually be a statement of *no effect* or *no change* or *no difference*. Examples are that a new drug does *not* reduce blood pressure; that changes in the reception arrangements at an outpatient clinic do *not* increase patients' satisfaction or reduce their anxiety; that there is *no* increase in patients' mobility after physiotherapy; and so on.

MEMO

The null hypothesis is usually one of no effect or no change or no difference, which we hope to be able to reject.

Anyway, suppose on the first day of its opening 20 people volunteered to become blood donors, and that 10 of these individuals were men and 10 were women. So the sample consists of 20 individuals drawn from the presenting population of potential blood donors, and the sample proportion of females is thus $p_{female} = 10/20 = 0.500$ (these data would be described as being *dichotomous*, because they can take only one of two values, male or female).

I think you would feel that such a sample result, being in fact identical with the hypothesised value, would oblige you to accept without much question that H_0 was true (and that Vlad was right – again). Now suppose instead that there were nine women and 11 men in the sample, giving $p_{female} = 9/20 = 0.450$. Would you still be inclined to accept H_0? Probably. What if there were eight women and 12 men giving a $p_{female} = 8/20 = 0.400$, would you still accept H_0? Maybe, maybe not. And if there were seven women giving a $p_{female} = 0.350$? Perhaps now we would entertain serious doubts as to whether π_{female} could equal 0.5, since a sample proportion of 0.350 is quite some distance from 0.5*.

* Interestingly enough, when I have posed this same question to groups of professional health carers in training sessions, or to health students whom I have taught, about 20% of them say they would reject H_0 when the number of females fell to nine or below (!), about 70% say they would reject H_0 when the number of women fell to eight or below, and about 10% say they would reject H_0 if the

Interesting as it is, this is all a bit hit and miss. Clearly, what we need is some sort of general *systematic* procedure to help us make a decision as to whether or not to accept or reject any null hypothesis. I am going to describe such a procedure in a moment, but first we need to think about what is called *error* in the context of hypothesis testing.

Types of Error

Suppose, following a gut feeling, we do decide to reject H_0 when $p_{female} = 0.350$, on the basis that this value is too far away from 0.5 for H_0 to be true. We could be making the *wrong* decision. It's entirely possible that such a value *could* occur by chance even if the true value of π_{female} was 0.5. It may not be very probable but it could still happen. Even if there was only *one* woman (or indeed no women) in the sample H_0 could still be true. The probability of this happening by chance might be very small but it is not impossible. In deciding to reject H_0 when in fact it's true, we are making what is known as a *type I error* (or sometimes as a *false positive* error).

In a similar way, we might accept H_0: $\pi_{female} = 0.5$ as being correct because the sample proportion, p_{female}, is 0.5, and again be mistaken. A sample value of 0.5 *could* occur by chance even if the true population proportion was, say, 0.45 or 0.6, or 0.9, and so on. In deciding to accept H_0 when it is false (and we should be rejecting it) we are committing what is known as a *type II error* (or sometimes as a *false negative* error).

MEMO

Rejecting H_0 when it is true is a type I (or false positive) error. Accepting H_0 when it is false is a type II (or false negative) error.

Statisticians conventionally set a maximum acceptable value for the probability of committing a type I error, of 0.05 or 0.01. This is known as the *significance level* of the hypothesis test and is denoted α (pronounced "alfa"). The probability of committing a type II error is usually denoted β (pronounced "beeta"). The analogy of a criminal trial is often quoted to illustrate the idea of hypothesis testing. The null hypothesis is that a person starts off being assumed innocent (not guilty). The prosecution offers evidence to try to disprove this assumption, the defence to support it. The jury weighs up the evidence and then either accepts the innocence of the accused and finds them not guilty (i.e. accepts the null hypothesis) or finds the evidence convincing enough to convict (rejecting the null hypothesis). In this analogy, a type I error amounts to finding an innocent person guilty, a type II error to returning a verdict of not guilty on somebody who did in fact commit the crime.

---

number of women fell to seven or below. A few students say they wouldn't reject H_0 unless the number of women fell to six or below. They are keen thereafter to see what a formal decision-making procedure suggests is the "correct" outcome.

MEMO

The probability of committing a type I error, rejecting the null hypothesis when it is in fact true, is also the significance level of the test, denoted α.

The Power of a Test

The *power of a test* is its ability to detect significant results, which may in some circumstances be small. Suppose we are testing a new drug which is designed to reduce diastolic blood pressure. The null hypothesis is that there is *no* significant effect. We carry out a trial which produces a *real* but small reduction in blood pressure, but because the effect is small the test does not detect it and we incorrectly accept H_0. We have committed a type II error, the probability of which is β, i.e. the probability of accepting H_0 (or not rejecting it) when it is false. The power of a test is defined as being equal to $(1 - \beta)$. Clearly we want tests with the power to detect significant, but small, effects or differences. To increase $(1 - \beta)$ means reducing β. But, unfortunately, it can be shown that decreasing β means increasing α (the probability of committing a type I error). There is a trade-off between the two. The only realistic way of increasing the power of any given test, without increasing α, is to increase the sample size. To determine an appropriate sample size it is necessary to specify the following four criteria:

- the significance level, α;
- the maximum acceptable chance of committing a type II error, β;
- the minimum size of the effect that would be considered significant (i.e. the minimum difference from the value specified in the null hypothesis which one would wish to detect);
- and, in the case of hypothesis testing about means, an estimate of population variance.

It is not possible in a book such as this to present any further detailed information about sample size, which can be quite complicated, other than that already given in Chapters 2 and 3, and interested (and disappointed) readers will have to refer to more advanced texts[1]. However, the EPI package does have a very useful facility which allows sample size calculations when the hypothesis being tested involves proportions. This is accessed through the command sequence:

Programs
Statcalc
Sample size & power

The screen then presents the following choice:

Population survey
Cohort or cross-sectional
Unmatched case-control

It is then only necessary to respond to questions on the screen relating to the four criteria referred to above to obtain a minimum sample size.

MEMO

The power of a test is its ability to detect possibly small but significant effects, i.e. to avoid accepting H_0 when it is not true (a type II error) and is equal to $(1 - \beta)$.

So far we've seen that whether we decide to accept or to reject the null hypothesis we might be making the wrong decision. But how do we make the decision in the first place?

Making the Decision: the *p*-Value

To decide whether or not to accept or reject H_0, we calculate what is known as the *p-value*. The *p*-value is the probability of getting a sample value as extreme as or more extreme than the one we *actually get*, when the null hypothesis is true. That is, getting a sample value that is far, or even further, away from the null hypothesis value. For example, suppose in our sample of 20 blood donors, there are five women (and 15 men). In this case the *p*-value is the probability of getting five or fewer women (or 15 or more women), when the null hypothesis, i.e. H_0: $\pi_{female} = 0.5$, is true.

How is the *p*-value calculated? Usually we would, if doing it by hand, have to make use of a *z* table or a *t* table or maybe a table of chi-squared values, depending on the problem we were dealing with. Fortunately, all of the computer programs we are considering in this book usually calculate the *p*-value for us, which makes life a lot easier. The decision is then as follows:

> If the *p*-value is less than the α value (the significance level of the test, i.e. the probability of committing a type I error) we reject H_0, otherwise we accept H_0.

So, for example, if the computer produces a *p*-value of 0.03 for some test, and the chosen significance level α is 0.05, then we could reject H_0. However, if the chosen significance level α was 0.01, we could not reject H_0.

If we are able to reject H_0, we say that *the hypothesis test is significant* at the 0.05 or 0.01 levels, depending on our chosen α value.

MEMO

The *p*-value is the probability of getting a sample value as extreme as or more extreme than (i.e. that far away from the null hypothesis value or even further) the one we actually got.

We can illustrate calculation of the *p*-value for our blood donating example by making use of the binomial distribution, which we have already made passing references to in our earlier discussion on estimation. We can use Minitab to calculate the binomial probabilities for us. To do this the values 0 to 20 should be entered into the first 20 rows of column c1, say, and then the mouse clicked on:

Calc
  Probability Distributions
    Binomial

Then set **Number of trials** to 20, **Probability of success** to 0.5 and **Input column** to **c1**. After clicking on **OK** the output shown in Figure 5.1 is obtained.

The data in Figure 5.1 show the probability of getting *by chance alone* all possible numbers of women, when we take a random sample of 20 individuals from a population containing *equal* numbers of men and women, i.e. when $\pi_{female} = 0.5$ (that is when H_0 is true)*.

Suppose in fact that on the first day we got five women out of the 20 individuals volunteering to give blood. We want to know whether this enables us to accept the null hypothesis or not. If we choose a significance level of $\alpha = 0.05$ we need to determine the *p*-value so that we can compare it with 0.05 and thus accept or reject H_0. To find the *p*-value we first have to determine the probability of getting *five or*

* The probability of getting 0 women or 1 woman is not actually zero, it is just too small to be captured with the first four decimal places.

```
MTB > PDF c1;
SUBC> Binomial 20.5.

Probability Density Function

Binomial with n = 20 and p = 0.500000

        x            P(X = x)

     0.00            0.0000
     1.00            0.0000
     2.00            0.0002
     3.00            0.0011
     4.00            0.0046
     5.00            0.0148
     6.00            0.0370
     7.00            0.0739
     8.00            0.1201
     9.00            0.1602
    10.00            0.1762
    11.00            0.1602
    12.00            0.1201
    13.00            0.0739
    14.00            0.0370
    15.00            0.0148
    16.00            0.0046
    17.00            0.0011
    18.00            0.0002
    19.00            0.0000
    20.00            0.0000
```

Figure 5.1: Using Minitab for binomial probabilities when $n = 20$ and $\pi = 0.5$

fewer women from this population if H_0: $\pi_{\text{females}} = 0.5$ is true. This is the probability of getting five women or four women or three women or two women or one woman or no women. This is illustrated in Figure 5.2. From Figure 5.1 this value is:

$$0.0148 + 0.0046 + 0.0011 + 0.0002 + 0.0000 + 0.0000 = 0.0207.$$

This is not quite the *p*-value, there is one more step we need to take.

The Mysterious Case of the Cat with Two Tails

The alternative hypothesis in the blood donating example was your belief that π_{female} is not equal to 0.5, i.e.

$$H_1: \pi_{\text{female}} \neq 0.5.$$

So if the null hypothesis is *not* true, π_{female} could be *either* smaller than 0.5 or larger than 0.5. When the alternative hypothesis is of this "not equal to" sort, the hypothesis test is described as being *two-tailed*. Suppose, however, that experience

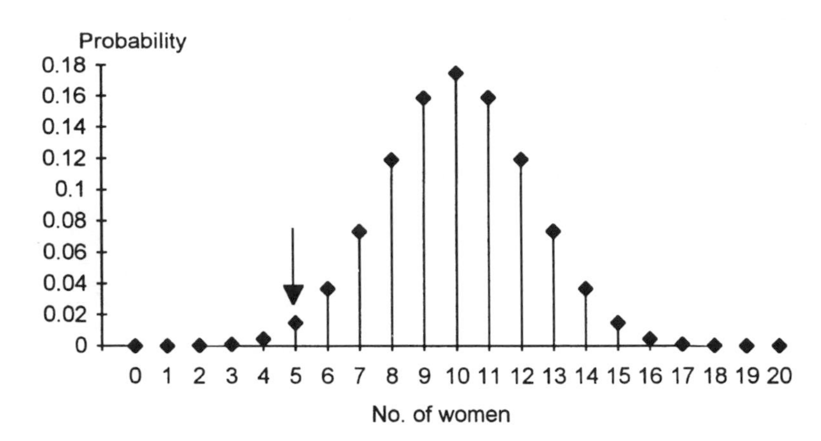

Figure 5.2: Probability of getting various numbers of women in a sample of 20 individuals volunteering to give blood when the population proportion of women is 0.5

had told us that usually more women come forward to donate blood than men. In this case you might have felt justified in defining the alternative hypothesis as:

$$H_1: \pi_{female} > 0.5.$$

In other words, if π_{female} is not 0.5 it is most likely to be *larger* than 0.5. On the other hand, experience might indicate that it is usually men rather than women who donate blood. In which case it might seem reasonable to define the alternative hypothesis as:

$$H_1: \pi_{female} < 0.5.$$

In other words, if π_{female} is not 0.5 it is most likely to be *smaller* than 0.5. When the alternative hypothesis is written in either of these two forms, it is described as *one-tailed*. As other authors have indicated[2], it is almost always safer to work with hypothesis tests of the two-tailed type. If you follow this advice you won't have to worry about what sort of hypothesis to use and in what circumstances, and the use of statistical tables, which often seem confusing if you have to decide between one- and two-tailed tests, will be easier.

Let us return now to our *p*-value. Since in this example we are working with a two-tailed test, we have to allow for the fact that the sample value may be as far away from the null hypothesis value of 0.5 (i.e. 10 women in the sample) in both the direction of fewer women (five or less), as *well* as in the direction of more women (15 or more women, to keep it symmetrical around 10 women). This means that we have to double the probability value we found above to arrive at the proper *p*-value. That is, when we get five women in the sample of 20 individuals, when $H_0: \pi_{females} = 0.5$ is true then the *p*-value is equal to $2 \times 0.0207 = 0.0414$.

We can now make our decision to accept or reject H_0. Remember, if the *p*-value is *less* than α we reject H_0, otherwise we accept H_0. In this case, the *p*-value is 0.0414. So if $\alpha = 0.05$ we would *reject* $H_0: \pi_{females} = 0.5$. You win the bet and Vlad has to buy the cakes. It would seem that, if we get only five women in the sample of 20, and we've chosen a significance level of 0.05, we must reject the null hypothesis that women make up half the would-be blood donors in this population. In these circumstances we would say that the *test is significant at 0.05*. We

It's the superior attitude that gets up my nose.

don't in fact know whether the true population proportion is less or more than 0.5, just that it is not equal to 0.5. We might use a bit of common sense, however, and say that with only five women out of a sample of 20, the proportion of women blood donors in the population is most likely to be less than a half.

Notice though that if we had set a significance level of $\alpha = 0.01$, we would instead have to *accept* the null hypothesis that $\pi_{\text{females}} = 0.5$, since now the p-value of 0.0414 is *greater* than an α of 0.01. In these circumstances we would say that the *test is not significant at 0.01*. To avoid buying the cakes you would have to ensure that you decide on a value of 0.05 for α beforehand! This would be considered to be *very* bad practice; the significance level should always be decided at the start of the hypothesis testing procedure.

MEMO

The hypothesis test decision rule is: reject the null hypothesis if the p-value is less than α (the significance level of the test), otherwise accept the null hypothesis.

To sum up the hypothesis testing procedure:

- *Step 1:* Set up the null and alternative hypotheses.
- *Step 2:* Decide on the significance level α (usually 0.05 or 0.01).

- *Step 3:* Get the relevant sample value (this might be a sample mean, or a sample median, or the difference between two sample means or proportions, and so on).
- *Step 4:* Determine the p-value, i.e. probability of getting this sample result (or one even worse) by chance when the null hypothesis is true. If you are doing the hypothesis test by hand this will usually mean looking up the probability in the appropriate table, e.g. a t table, or a binomial table, or a chi-squared table, or whatever.
- *Step 5:* If this p-value is less than 0.05 or 0.01, whichever significance level has been set, reject the null hypothesis, otherwise accept the null hypothesis.

As you will see shortly we can use computer packages to perform the binomial test.

NOMINAL DATA: ONE SAMPLE

What you have read so far summarises the most important ideas underlying the concept of hypothesis testing. In the rest of this chapter I will outline only briefly the main features of the most important tests. I'll begin by looking at tests we can use with nominal data, then follow this in subsequent chapters by describing tests we can use with ordinal data, and conclude with metric data tests. It seems appropriate to start with the binomial test.

The Binomial Test

- *Used with:* nominal *dichotomous* data in one sample.
- *Used to test:* whether the *proportion* of subjects in a population with some attribute or quality is equal to some hypothesised value; for example, that the proportion of all the patients registered with a health practice who smoke is 0.1 (i.e. 10%).

Brief Description of the Binomial Test

The rationale behind the binomial test has been outlined above in the blood donor example, but in essence if a dichotomous or binary variable is sampled under conditions which satisfy the requirements of the binomial distribution (repeated independent trials on dichotomous variables with fixed probability of success), then we can use the binomial probabilities to test the null hypothesis that the population proportion is equal to some specified value (for example 0.5).

We can use both SPSS and Minitab to apply the binomial test to the blood donors example described above.

Using SPSS to Perform the Binomial Test

To use SPSS, enter the raw data, five 1s and fifteen 0s in column 1, say, of the data sheet (which I named "nowomen"). Then click on the following commands:

Statistics
  Nonparametric Tests

Binomial
  Select c1 (in the "Test Variable List" box)
    ⊙ **Get from data** (in the "Define dichotomy box")
      OK

This accepts the default "Test Proportion" of 0.5 (this being the null hypothesis value in this example, but the test proportion can of course be set anywhere between 0 and 1 as required). The output shown in Figure 5.3 should then appear. The p-value of 0.0414 is the same as we obtained by hand above, enabling us to reject H_0.

```
Binomial Test

NOWOMEN

Cases
                     Test Prop. = 0.5000
    5 = 1            Obs. Prop. = 0.2500
   15 = 0
Exact Binomial
   20 Total          2-Tailed P = 0.0414
```

Figure 5.3: Using SPSS for the binomial probability when the sample of 20 blood donors contains five women and the null hypothesis is that $\pi_{\text{females}} = 0.5$

Using Minitab to Perform the Binomial Test

Minitab does not actually do the binomial test directly. However, it can be used to calculate the probability of getting any number of successes (x), or a fewer number, in any given number of trials (n). It does this by calculating what it calls the *cumulative probability density function* (CDF). This language is quite technical, but to do this is quite simple. For example, in the blood donors example, first enter the value 5 (for five women) in column c1, say, then follow the command sequence:

Calc
  Probability Distributions
    Binomial
      ⊙ **Cumulative probability**
        20 (in "Number of trials" box)
          0.5 (in "Probability of success" box)
            ⊙ **Input column**
              c1
                OK

The output shown in Figure 5.4 is then produced. The probability of getting five or fewer women is given as 0.0207 (the same result as calculated by hand above), which we know we have to double to get the p-value of 0.0414, because we have to include the probability of getting 15 or more women. The result is the same as by hand and from SPSS. The Minitab procedure is quicker than SPSS in that the raw scores do not first have to be entered (of which there may be *many* more than 20).

```
MTB > CDF ''No.Women'';
SUBC> Binomial 20 0.5.

Cumulative Distribution Function

Binomial with n = 20 and p = 0.500000

    x        P(X <= x)
  5.00       0.0207
```

Figure 5.4: Using Minitab for the binomial probability of getting five or fewer women ("successes") in 20 trials when the population proportion is 0.5

The Chi-squared (χ^2) Goodness-of-fit Test*

- *Used with:* nominal data in a single sample which can be classified into two or more categories.
- *Used to test:* whether the actual or *observed* proportion of subjects in each category in a sample corresponds with the hypothesised or *expected* proportions. One very common application is to test whether the proportions in categories are all the *same*.

Brief Description of the Chi-squared Goodness-of-fit Test

It will be easiest to illustrate the chi-squared goodness-of-fit test with an example. Consider the data in the first row of Table 5.1 which shows the percentage of individuals (cases) who were *observed* to have quit smoking in a case–control study[3], using one of four different nicotine replacement therapies (NRT). The percentages quitting were: gum (18%), patches (21%), nasal spray (26%), inhaler (15%). Suppose the null hypothesis is that all methods are *equally effective.* If this hypothesis is true we would have *expected* each method to have achieved a quit rate equal to the average of these four, i.e. a quit rate equal to (18 + 21 + 26 + 15)/4 = 20%. The second row of Table 5.1 shows these expected values. The observed values are labelled "*O*" and the expected values "*E*".

Table 5.1: Calculations for the chi-squared goodness-of-fit test for the quitting smoking example

	GUM	PATCHES	NASAL SPRAY	INHALER
Observed values O	18	21	26	15
Expected values E	20	20	20	20
$(O-E)$	−2	1	6	−5
$(O-E)^2$	4	1	36	25
$(O-E)^2/E$	4/20 = 0.25	1/20 = 0.05	36/20 = 1.8	25/20 = 1.25

* "Chi" is pronounced as in the first syllable of *Cairo*.

The rationale underlying the chi-squared test is that if the observed (O) proportion in each category is the same as that which we expected (E) to find when the null hypothesis is true, then the value of each $(O - E)$ term, i.e. the difference between them, should be approximately zero (only approximately and not exactly so because of chance variation in the sample). It follows that the sum of all the $(O - E)$ values should also be approximately zero, and if it is we can accept the null hypothesis.

Unfortunately, the sum of the $(O - E)$ values is *always* zero because of the cancelling out of positive and negative values. It's the same problem we have with the sum of the $(x - \bar{x})$ values in the standard deviation calculation. We get round the s.d. problem by squaring the $(x - \bar{x})$ values and we can do the same in this situation, by squaring each $(O - E)$ value. Clearly, if each $(O - E)$ value is zero then the square is also going to be zero and the sum of the squared values will be zero. We finally divide each squared $(O - E)$ value by E to adjust or standardise for the size of each $(O - E)$ term before summing.

We can summarise the chi-squared procedure thus:

- *Step 1:* Determine the expected values in each category assuming the null hypothesis to be true (the E values).
- *Step 2:* Take a sample and get the observed values in each category (the O values).
- *Step 3:* Calculate the $(O - E)$ value for each category.
- *Step 4:* Square each $(O - E)$ value.
- *Step 5:* Divide each $(O - E)^2$ value by the E value for that category (to standardise).
- *Step 6:* Sum all the $\frac{(O - E)^2}{E}$ values to give the value of the term

$$\sum \frac{(O - E)^2}{E} \quad \text{or} \quad \Sigma\, [(O - E)^2/E].$$

Given that the null hypothesis is true the value of this last term should be approximately zero. We know it won't be *exactly* zero because of *chance* variation in the sample, so the crucial question is: how far away from zero does the value have to be before the probability of it happening by chance alone (when H_0 is true) is too small for us to accept the null hypothesis? Or putting it another way, what is the probability of getting the value we got for $\sum \frac{(O - E)^2}{E}$ or one even further away from 0, and is this probability less than α, the significance level we have chosen for this test? This is where the chi-squared distribution comes in handy. It so happens that the term $\sum \frac{(O - E)^2}{E}$ has the same shaped distribution (is distributed) as the chi-squared statistic*. So we can use the values from

* Recall that when we say something "is distributed as" something, we mean that if we were to take many samples from this population and for each sample calculate the value of $\sum \frac{(O - E)^2}{E}$ and then plot all of these values, the shape of the resulting curve would be the same as the shape of the chi-squared curve.

the chi-squared table to determine the probabilities of getting various values of $\sum \frac{(O-E)^2}{E}$ and decide whether these values are far enough away from zero for us to be able to reject the null hypothesis.

The third row of Table 5.1 shows the values of the $(O - E)$ terms, row four the values of $(O-E)^2$, and row five the values of $\frac{(O-E)^2}{E}$. By summing the values in this last row we get a value for $\sum \frac{(O-E)^2}{E} = 3.35$. This value isn't zero, but is it far enough away from zero for us to conclude that H_0 is probably not true? The decision rule is:

If $\sum \frac{(O-E)^2}{E} > \chi^2$ then reject H_0

Otherwise accept H_0.

A complete chi-squared table is given in Table A5 of the Appendix, but a small section is shown in Table 5.2.

Table 5.2: Values of the chi-squared distribution: If a variable X is distributed as chi-squared (χ^2), then α is the probability that $X \geq \chi^2$

DEGREES OF FREEDOM	α	
	0.05	*0.01*
1	3.84	6.63
2	5.99	9.21
3	**7.81**	11.34
4	9.49	13.28
5	11.07	15.09

The first column headed degrees of freedom (d.f.) indicates which row we must use. In a single sample case the d.f. are the number of categories minus 1. In this example there are four categories (gum, patches, nasal spray and inhaler) so the d.f. = 4 – 1 = 3, and we need therefore to use row three. If we set an $\alpha = 0.05$, the critical value is thus 7.81. Table 5.2 tells us that values of chi-squared greater than 7.81 will occur with a probability of less than 0.05 by chance alone. Values below 7.81 are less rare. Thus if the value of $\sum \frac{(O-E)^2}{E}$ is greater than 7.81 we can say that the value is too far away from zero to have occurred merely by chance and we can therefore reject H_0. In this example 3.35 is *less* than 7.81 so we accept H_0. The value is close enough to zero to have occurred by chance with a probability not rare enough for us to question the truth of the null hypothesis. We conclude therefore that the observed proportions of smoking quit rates are close enough to the expected values for us to believe the null hypothesis that all four methods of helping people quit smoking are equally effective.

An Example From Practice

Researchers investigating the use of a self-help therapeutic manual for managing bulima nervosa used a controlled trial design.[4] The subjects were randomised to three groups: those using the manual; those receiving cognitive therapy; and those left on the waiting list. The proportions of subjects abstaining from a number of bulimic behaviours (binge eating, vomiting, etc.) after eight weeks was tested to see if it was the same in each of the three groups, using the chi-squared goodness-of-fit test. The number of degrees of freedom used was 3 − 1 = 2, and the results quoted were:

Binge eating	$\chi^2 = 1.79$, $p = 0.41$
Vomiting	$\chi^2 = 0.69$, $p = 0.71$
Other weight-control behaviours	$\chi^2 = 6.29$, $p \leq 0.05$.

These results indicate that there is no difference between the three groups in terms of binge eating or vomiting, but they do differ in terms of "other weight-control behaviours".

Using a Computer for the One-sample Chi-squared Goodness-of-fit Test

Minitab does not calculate chi-squared directly and requires that the value of $\sum \frac{(O-E)^2}{E}$ is first calculated followed by the *p*-value. This requires typing in a number of Minitab commands. This is clearly not very convenient. Details can be found in the Minitab Reference Manual.

As an example of the use of SPSS to perform a one-sample chi-squared goodness-of-fit test we can use data taken from an investigation into a comparison of the mortality from natural causes of individuals who have previously attempted suicide[5]. This differs from the previous example on NRT therapies in that in this case the expected values are not all equal. SPSS requires that the *raw* data be entered into the datasheet. The original data record the cause of death, by one of eight categories, of a total of 106 individuals who had previously attempted suicide in Copenhagen. The 106 observations are entered in column 1 of the SPSS datasheet and labelled "cause". To make the output more intelligible, the numeric data are coded into the eight alphabetic disease categories using **Data, Define Variables, ⊙ Labels** (Value = **1**, Value Label = **Neoplasm;** Value = **2**, Value Label = **Endocrine**; etc.). The following commands are then used:

Statistics
  Nonparametric Tests
    Chi-Square
      ⊙ Get from data (default)
      Expected values

The following expected values in the eight mortality categories provided by the authors are then entered in the **Value** box, clicking on the **Add** button after each value:

18.93 0.99 0.51 25.7 3.76 1.96 0.62 1.33

These expected values are those experienced by the general population. The authors' intention is to determine whether the pattern of natural deaths among previously-attempted suicides matches that in the general population (i.e. whether or not there is a good fit between the two patterns). The authors' null hypothesis is that mortality patterns in the general population are the same as that in those who have previously attempted suicide.

The SPSS output is shown in Figure 5.5 where a chi-squared of 122.9981 is calculated with a p-value less than 0.0000, i.e. the test is highly significant.

In other words, there is a significant difference in the pattern of deaths between the subjects and what would be expected in the general population. Notice that the expected values used by SPSS to calculate chi-squared and given in Figure 5.5 are *twice* those entered in the chi-squared dialogue box. This is because the expected values supplied by the authors add up only to 53, half of the number of sample observations. Necessarily the total number of expected and observed values must be the same.

```
SPSS for MS WINDOWS Release 6.

Chi-Square Test

CAUSE     Cause

                              Cases
         Category            observed        Expected        Residual

Neoplasm          1.00          32             37.30           -5.30
Endocrine         2.00           2              1.95             .05
Nervous           3.00           2              1.00            1.00
Circulatory       4.00          33             50.64          -17.64
Respiratory       5.00           8              7.41             .59
Digestive         6.00           8              3.86            4.14
Alcohol           7.00          12              1.22           10.78
Other             8.00           9              2.62            6.38
                               ---
              Total            106

Warning—Chi-Square statistic is questionable here.
5 cells have expected frequencies less than 5.
Minimum expected cell frequency is    1.0

Chi-square       D.F.       Significance
122.9981         7          0.0000
```

Figure 5.5: SPSS output from the chi-squared one-sample goodness-of-fit test, for the difference in cause of death between the general population of Denmark and 106 subjects who had previously attempted suicide

Notice that the output also carries a warning about the small expected value of five out of the eight cells. It is unwise to depend on the reliability of a chi-squared test when more than 20% of expected cell values are less than 5, or any one cell has an expected value less than 1*. This is because for small expected cell values,

* When there are only two categories, *every* cell must have an expected value of at least 5.

the approximation of $\sum \frac{(O-E)^2}{E}$ to the chi-squared distribution is not particularly good. One way to overcome this problem is to amalgamate categories if this can be done meaningfully. Failing this the binomial two-category test may have to be used.

HYPOTHESIS TESTS WITH TWO MATCHED SAMPLES

The McNemar Test

- *Used with:* nominal (or ordinal) dichotomous data.
- *Used to test:* whether or not there is a relationship between two variables, i.e. whether or not the two variables are independent. Often used for before-and-after, self-matching samples. For example, to assess the significance of *change* before and after some intervention. (Note that a chi-squared test *cannot* be used with two *matched* samples.)

Brief Description of the McNemar Test

The McNemar test determines independence (i.e. no relationship) between the two variables in question by testing whether or not the proportion of cases in each of the two categories in one group is the same as the proportion of cases in the two categories in the second group. For example, if we ask visitors to a geriatric ward whether or not ("yes" or "no") they like the decor and general ambience of the ward, some will say they do, others that they do not. The ward is now redecorated, furniture moved around, etc. and the same visitors are asked again for their opinion. Once again some will like it and some not. Some will have changed their mind, some will still be of the same opinion as before. Clearly the two samples, before and after, are matched. The raw data take the form shown in Table 5.3(a) (only the first few rows are shown), which can also be formulated as a 2×2 table, as in Table 5.3(b). Note that a total of $(a + d)$ visitors feel the same before as they do after and a total of $(b + c)$ visitors have changed their mind.

It can be shown that the term $(b - c)^2/(b + c)$ is distributed as chi-squared with one degree of freedom, so the test is based *only* on those individuals who *change* their opinion. The value of $(b - c)^2/(b + c)$ is calculated and compared with the value of chi-squared with one degree of freedom (i.e. using row one) from a chi-squared table. If $(b - c)^2/(b + c)$ is greater than the chi-squared table value we reject the null hypothesis, otherwise we accept the null hypothesis.

Example from Practice

Investigators used the McNemar test to determine whether there had been a significant reduction in the use of nine health and social service facilities available to a sample of 51 mental health patients between one year and four years after their discharge from hospital.[6] Use was defined as at least one contact with the relevant service in the three-month period prior to interview.

McNemar is appropriate here since these are matched (same individuals) samples, in nominal categories. The researchers found significant reductions in the use of only three

Table 5.3: Visitors' opinions on decor and ambience of geriatric ward, before and after revamp

(a) Raw data in response to question to visitors, "Do you like the decor and ambience of the ward?", before and after revamping (first few rows only)

VISITOR	BEFORE	AFTER
1	Yes	Yes
2	No	Yes
3	No	Yes
4	No	No
5	No	Yes
6	Yes	Yes
.	.	.
.	.	.
.	.	.

(b) Data expressed in the form of a 2 × 2 table

		AFTER	
		Yes	*No*
BEFORE	*Yes*	*a*	*b*
	No	*c*	*d*

out of the nine services available. For example, the number of patients contacting their GP at one year was 34 and at four years was 26, a non-significant reduction (p-value $\geq$ 0.06). However, the number of contacts with a social worker was significantly down from 17 to 7 (p-value $\leq$ 0.03).

Using a Computer to Do the McNemar Test

Let us use the fictitious example above with a sample of 37 visitors to the geriatric ward and with $a = 8$, $b = 7$, $c = 19$ and $d = 3$. The null hypothesis is that the same proportion change their opinions in one direction as in the other, i.e. there is no substantive change in visitors' overall opinion from before redecoration to after.

Using SPSS for the McNemar Test

Two columns of data are entered into the data sheet. The first, labelled "Before", contains the views of the visitors before the redecoration of the ward, the second, labelled "After", their views after redecoration. "Yes" is coded 1 and "No" is coded 0 in both columns, using the **Labels** button in the **Define Variables** dialogue box (accessible from the **Data** menu). To obtain the McNemar test, click on the command sequence:

Statistics
  Nonparametric Tests
    2 Related Samples
      Select Before and After (as current selections)
        ☐ **Wilcoxon** (to remove this as default test)

☒ **McNemar** (to include)
OK

If there are fewer than 25 pairs of observations SPSS uses the more exact binomial distribution. In this case there are 37 values so chi-squared is used. The output shown in Figure 5.6 is produced. The value of $(b - c)^2/(b + c)$ is 4.6538.

```
McNemar Test

AFTER             After
with BEFORE       Before

                    BEFORE

                    1         0         Cases     37

            0       7         3         Chi-squared     4.6538
AFTER
            1       8        19         Significance    0.0310
```

Figure 5.6: SPSS output from the McNemar test on attitude of visitors to ambience in a geriatric ward

The p-value is given as "Significance 0.0310". Since 0.0310 is less than $\alpha = 0.05$ (our chosen level of significance) we are able to *reject* the null hypothesis of no change between visitors' opinions about the ward before and after the revamp. Notice that had we chosen a level of significance $\alpha = 0.01$, we would *not* have been able to reject the null hypothesis since now the p-value of 0.0310 is not less than $\alpha = 0.01$.

Using Excel for the McNemar Test

The standard version of Excel does not include the McNemar test. However, if the Astute add-in is available this can be used. Astute, unlike SPSS, does not require the original data to be entered but allows the 2 × 2 *contingency table* to be entered directly into the spreadsheet. A contingency table is a table in which the columns represent the different samples or groups and the rows represent the categories of the variables concerned. All that is required for the test to be performed is that the data are first selected, and the McNemar button on the Astute toolbar is selected. The input and output ranges are then requested in the dialogue box and OK clicked. As with SPSS, Excel gives the value of the chi-squared statistic and the two-tailed p-value. Space restrictions do not permit me to provide an example.

Using Minitab for the McNemar Test

Release 10 of Minitab does not include the McNemar test.

MEMO

A contingency table is a table in which the columns represent the different samples or groups and the rows represent the categories of the variables concerned.

HYPOTHESIS TESTS WITH TWO INDEPENDENT SAMPLES

The test most often used with two *independent* samples or groups measured at the nominal level is the chi-squared (χ^2) test. Each group can have the data arranged into two or more categories. For example, in a study into the relationship between childhood abuse and later-life depression in women, the two groups might be women abused and women not abused in childhood, with maybe three categories in each group, chronic depression, single episode depression, and no depression. This particular example would be known as a 3 by 2, or 3 × 2 example (number of categories × number of groups). I will describe the chi-squared test by starting with its simplest form, the 2 × 2 test.

The Chi-squared 2 × 2 Test

- *Used with:* two *independent* groups or samples each with nominal (or ordinal) data in two categories.
- *Used to test:* whether or not there is a relationship between two variables, i.e. whether or not the two variables are independent.

Brief Description of the Chi-squared 2 × 2 Test

Chi-squared tests independence (no relationship) between the two variables by testing whether or not the proportion of cases in each of the two categories in one group is the same as the proportion of cases in the two categories in the second group. I will illustrate the general idea with an example drawn from practice.

An Example from Practice

In a study into the prevalence of HIV among adult patients attending an accident and emergency department in an inner-city area,[7] researchers studied a sample of 918 patients. They found that, among 263 patients who were foreign visitors, 9 were HIV positive. In this example the two groups or samples are those who are and those who are not HIV+, and the two categories are foreign visitor or not foreign visitor. The chi-squared 2 × 2 test was used to find out if there was a relationship between the variable "being a foreign visitor" and the variable "being HIV positive", in this population of A&E patients. In other words, is there any connection between the two variables? The hypotheses were thus:

H_0: HIV status and foreign visitor status are independent (no relationship between variables)

H_1: there is such a relationship.

The chi-squared test works by comparing the proportions in each of the two categories in the two groups. If these proportions are the same we can accept the null hypothesis and conclude that there is no relationship between the two variables (i.e. they are independent and therefore whether the subject is a foreign visitor or not has no bearing on whether they are likely to be HIV positive). If the proportions differ then we can reject the null hypothesis and conclude that there is a relationship between the two variables.

We would normally start this test with two columns of raw data as in Table 5.4(a). The test is referred to as a 2 × 2 because the raw data can be recast into a 2 × 2 contingency table, as in Table 5.4(b).

Table 5.4: Tables for the 2 × 2 chi-squared test on HIV status of A&E patients and whether foreign or not

(a) Raw data (Y = foreign visitor; N = not foreign visitor)

GROUPS	
HIV+	*Not HIV+*
N	N
Y	Y
N	N
N	N
N	Y
N	N
Y	N
.	.
.	.

(b) The contingency table

		HIV STATUS		Totals
		Positive	Negative	
FOREIGN VISITOR	Yes	*a* (9)	*b* (254)	263
	No	*c* (3)	*d* (652)	655
	Totals	12	906	918

The principle underlying the 2 × 2 chi-squared test is the same as for the chi-squared goodness-of-fit test discussed above. The proportion of cases *expected* (E) in each category when the null hypothesis is true is compared with the *observed* (O) proportion. If the value of the term $\sum \frac{(O-E)^2}{E}$ exceeds the value of chi-squared from Table A5 we reject H_0, otherwise we accept H_0. If the calculation has to be done by hand, then from the contingency table the expected value of any cell is easily found by multiplying the total of the row in which the cell lies by the total of its column and dividing the result by sample size n (this applies to contingency tables of any size). For example, using the contingency table in Table 5.4(b), the expected value for cell a (HIV+ and foreign visitor) if H_0 is true, is $(12 \times 263)/918 = 3.438$, compared with an observed value of 9.

Alternatively, rather than working out the expected values and substituting in the above summation expression, the following short-cut formula may be used (but only of course in the 2 × 2 case):

$$\Sigma \frac{(O-E)^2}{E} = \frac{(ac-bd)^2 \times n}{(a+b)(c+d)(a+c)(b+d)} .$$

Using a Computer to Do a Chi-squared 2 × 2 Test

Using Minitab for the Chi-squared 2 × 2 Test

To use Minitab for the chi-squared 2 × 2 test in the HIV/foreign visitor example above, first enter the contingency table values from Table 5.4(b) into two columns of the worksheet (9 and 3 in, say, column c1, and 254 and 652 in c2). Then click on the following sequence:

Stat
  Tables
    Chisquare
      Select c1
        Select c2
          OK

If only the raw data are available you should choose the **Stat, Tables, Cross Tabulation, Chisquare,** which will construct the contingency table before doing the chi-squared test. The output from the above procedure is shown in Figure 5.7.

```
MTB > ChiSquare ''HIV+'' ''HIV-''.

Chi-Square Test

Expected counts are printed below observed counts

            HIV+      HIV-     Total
    1          9       254       263
            3.44    259.56
    2          3       652       655
            8.56    646.44
 Total        12       906       918

ChiSq = 8.999 + 0.119 +
        3.613 + 0.048 = 12.779

df = 1, p = 0.000

1 cell with expected counts less than 5.0
```

Figure 5.7: Minitab output for the chi-squared 2 × 2 test on HIV status of A&E patients

The expected values are calculated and shown in the output below the observed values (notice that Minitab calculates an expected value for cell *a* of 3.44, the same value we got by hand above). Minitab calculates a value for $\sum \frac{(O-E)^2}{E}$ of 12.779

with a p-value of 0.000. So the test is highly significant* and we would reject H_0. There *does* appear to be a relationship between the patients' HIV status and whether or not they are a foreign visitor or not. Notice that the Minitab output carries the warning "one cell with expected counts less than 5". As we noted above, as a general rule it is unwise to use the chi-squared 2 × 2 test if any of the expected values fall below 5, because in these circumstances $\sum\frac{(O-E)^2}{E}$ cannot be relied upon to have a chi-squared distribution. In which case, it might be advisable instead to use Fisher's Exact test discussed below (although some statisticians maintain that Fisher's test is too conservative and that chi-squared is still applicable even for small samples[8]).

Using SPSS for the Chi-squared 2 × 2 Test

Although SPSS can be used to perform the chi-squared 2 × 2 test, the program does not allow the user to enter the contingency table as Minitab does, but unfortunately requires that the columns of *raw* data be entered (as it does with the chi-squared goodness-of-fit test above). Since this process will be very time consuming for largish samples, this is a weakness in the SPSS program. I will not therefore use the HIV example to illustrate the chi-squared 2 × 2 test using SPSS since this would require entering 918 observations, and life is too short!

Instead I will use fictitious data relating to a "did not arrive" (DNA) problem affecting psychiatric outpatient clinic appointments. This has been investigated elsewhere[9] and it seems that patients who have a depressive illness are more likely not to turn up than patients with other categories of mental illness. The null hypothesis is that whether a patient turns up for an appointment or not has no relation with (i.e. is independent of) whether their mental illness is depressive or not. The alternative hypothesis is that this is not true, and that there is a relationship between the two. Suppose we have the raw data in Table 5.5.

Table 5.5: Raw data on the number of patients arriving and not arriving at a psychiatric outpatient clinic by whether mental illness is of the depressive type or not

PATIENT	DEPRESSED	ARRIVED FOR APPOINTMENT	PATIENT	DEPRESSED	ARRIVED FOR APPOINTMENT
1	Y	Y	16	Y	N
2	Y	N	17	N	Y
3	N	Y	18	N	Y
4	N	Y	19	N	N
5	Y	N	20	Y	Y
6	N	N	21	Y	N
7	N	Y	22	N	Y
8	Y	Y	23	Y	N
9	Y	N	24	Y	N
10	N	N	25	N	Y
11	N	Y	26	N	Y
12	N	Y	27	Y	N
13	Y	N	28	Y	N
14	Y	Y	29	N	Y
15	Y	Y	30	Y	N

* The p-value is of course not 0, it's just that Minitab only gives the first three decimal places, so it must be 0.0009999 or less.

The raw data are entered into columns c1 (depressive illness or not) and c2 (arrived for outpatient appointment or not) of the SPSS data sheet and are coded, using the **Data, Define Variables, Type, Labels** commands, as 1 = Yes and 0 = No. By clicking on the following SPSS commands, the output shown in Figure 5.8 is produced:

```
SPSS for MS WINDOWS Release 6.

ARRIVED Arrived by DEPRESSD Depressed

                          DEPRESSD
                          No          Yes
                           0           1         Row totals (%)
  ARRIVED
                 0         3          10           13 (43.3)
                No
                 1        12           5           17 (56.7)
               Yes
  Column                  15          15           30 (100.0)
 total (%)              (50.0)      (50.0)

Chi-Square                Value        DF        Significance
Pearson                  6.65158       1           0.00991
Continuity               4.88688       1           0.02706
Correction
Likelihood Ratio         6.94641       1           0.00840

Minimum Expected Frequency — 6.500

Number of Missing Observations: 0
```

Figure 5.8: SPSS output for the outpatient DNA example

Statistics
  Summarise
    Crosstabs
      Select c1 (for the Rows box)
        Select c2 (for the Columns box)

Statistics

☒ **Chisquare**
Continue
OK

SPSS calculates a value for $\sum \frac{(O-E)^2}{E}$ of 6.65158, which it labels *Pearson* (the chi-squared statistics in the form we are using it was developed originally by a statistician named Pearson). The *p*-value is given as 0.00991, which is less than either 0.05 or 0.01 (just). We would thus reject H_0, and conclude that the likelihood of a patient arriving for an outpatient appointment *is* related to whether or not the patient's illness is depressive.

SPSS also calculates a value for chi-squared with a *continuity correction* (known as Yates' correction for continuity) of 4.88688 with a p-value of 0.02706. The continuity correction may be included because a continuous distribution (chi-squared) is being approximated by a discrete distribution, $\sum \frac{(O-E)^2}{E}$. Statisticians, however, argue over the merits of including this correction. Notice that, with the correction, the result is still significant at 0.05 but no longer significant at 0.01. (Remember that these are fictitious figures.)

Using Excel for the Chi-squared 2 × 2 Test

The standard version of Excel does not include the chi-squared test. However, the Astute add-in does allow this test. Excel, unlike SPSS, does not require the original data to be entered but allows the 2 × 2 *contingency table* to be entered directly. As an example we can apply the chi-squared test to the foreign visitor/HIV status contingency table data in Table 5.4(b). The four contingency table values should be entered into columns A and B, say, of the Excel spreadsheet. The easiest way to enter two columns of data into the spreadsheet with Astute is to select the first column by clicking on A, and then click on B while holding down the **CTRL** key. Then click on the chi-squared button on the Astute toolbar and on **OK** in the dialogue box. The output from Excel is shown in Figure 5.9.

```
Expected Numbers

              A          B
   1:        3.4       259.6
   2:        8.6       646.4

  Chi-square:      12.7790         10.5848
      df:          1               1
      p:           0.0004          0.0011
```

Figure 5.9: Output from the Excel Astute add-in chi-squared 2 × 2 test for the foreign visitor/HIV status data in Table 5.4

As with SPSS, Excel gives the value of the chi-squared statistic and the two-tailed p-value, without and with Yates' correction applied. The uncorrected p-value is given as 0.0004 and the corrected as 0.0011. Both values confirm rejection of the null hypothesis of no relation between foreign visitor status and HIV+ status.

Using EPI for the Chi-squared 2 × 2 Test

In many ways, EPI is the easiest computer program to use for the chi-squared 2 × 2 test. The required command sequence is:

Programs
　Statcalc
　　Tables

The screen will then show a blank 2 × 2 table with the columns labelled "Disease + or –" and the rows labelled "Exposure + or –". To use EPI with the HIV status example discussed above, 9 should now be typed and the return key pressed, followed in turn by the values 254, 3 and 652 (pressing the return key after each). Pressing the return key once more will produce the output shown in Figure 5.10. EPI gives an uncorrected value of chi-squared of 12.78, the same as Minitab, with a p-value of 0.00035. Yates' corrected value is 10.58, with $p = 0.00114$.

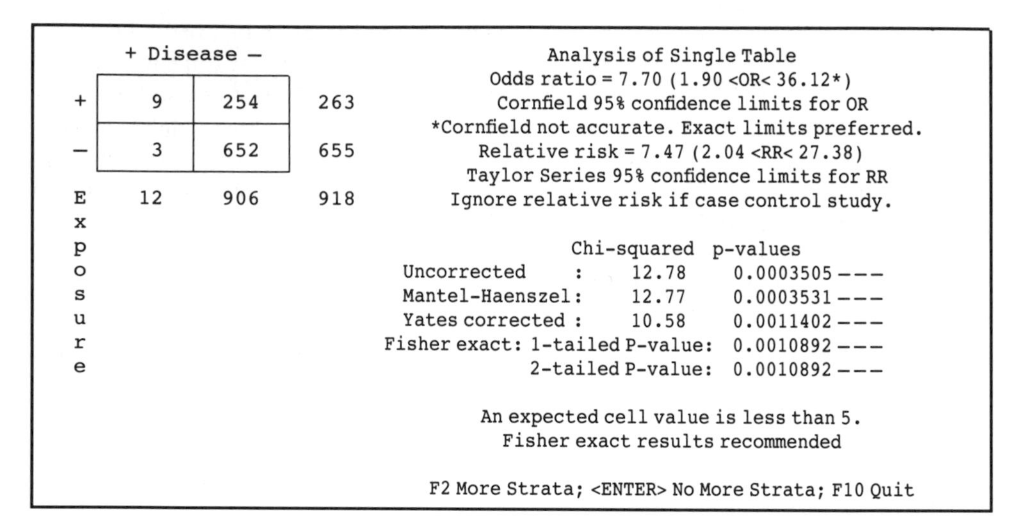

```
      + Disease –                          Analysis of Single Table
                                       Odds ratio = 7.70 (1.90 <OR< 36.12*)
 +  |  9  | 254 |   263                Cornfield 95% confidence limits for OR
                                    *Cornfield not accurate. Exact limits preferred.
 –  |  3  | 652 |   655                Relative risk = 7.47 (2.04 <RR< 27.38)
                                       Taylor Series 95% confidence limits for RR
 E    12    906     918                Ignore relative risk if case control study.
 x
 p                                          Chi-squared   p-values
 o                             Uncorrected     :   12.78     0.0003505 ---
 s                             Mantel-Haenszel:    12.77     0.0003531 ---
 u                             Yates corrected :   10.58     0.0011402 ---
 r                           Fisher exact: 1-tailed P-value: 0.0010892 ---
 e                                         2-tailed P-value: 0.0010892 ---

                                      An expected cell value is less than 5.
                                         Fisher exact results recommended

                                 F2 More Strata; <ENTER> No More Strata; F10 Quit
```

Figure 5.10: Output from EPI for the chi-squared 2 × 2 test on HIV status of A&E patients

The Relationship Between a Hypothesis Test Result and the Corresponding Confidence Interval

In the example above, the sample proportions of HIV+ and HIV– patients who are foreign visitors are 9/12 = 75% and 254/906 = 28% respectively. If we use CIA (as described in Chapter 3) to calculate a 95% confidence interval for the difference in the corresponding population proportions we get a confidence interval of (22.3% to 71.6%). This interval does *not* include zero, so we are able to conclude that there is a significant difference between the two proportions, and we can be 95% confident that this difference is somewhere between 22.3% and 71.6%. The result of the hypothesis test was also that there is a significant difference between the two groups in the proportions in each category (remember that the chi-squared test compares category proportions). *This is an important general result.* Whenever we perform a hypothesis test of the difference in two population parameters and are able to *reject* the null hypothesis, the corresponding confidence interval will *not* include zero. Conversely, if the confidence interval *does* include zero, we will *not* be able to reject the null hypothesis. So the confidence interval not only provides information about the likely size of any significant difference between two population parameters, it can also be used to perform the corresponding hypothesis test. And that's not all.

When there is a significant difference in the two population parameters, the confidence interval tells us something about the size of this likely difference; for example, somewhere between 22.3% and 71.6% in the example above. The hypothesis test does not provide that extra information, it merely indicates that the difference between the parameters is significant. This is why it is *preferable*, whenever it is practicable, to calculate confidence intervals rather than perform hypothesis tests.

Chi-squared Test for Tables Greater Than 2 × 2

The methods described above for the 2 × 2 chi-squared test can be applied equally to larger tables. All that is required is that the extra columns of data be entered into the datasheet. Calculation by hand takes longer since we can't use the short-cut formula for tables bigger than 2 × 2, but expected values are still calculated in the same way (row total × column total divided by n). Table 5.6, for example, is a 2 × 3 contingency table taken from a study into depression in adult women and their abuse in childhood[10].

Table 5.6: Example of a 2 × 3 contingency table for the chi-squared test, abuse in childhood by type of depression

ABUSE IN CHILDHOOD	TYPE OF DEPRESSION		
	Chronic recurrent	*Single episode*	*No depression*
Yes	14	3	14
No	9	13	48

The null hypothesis is that there is no relationship in the population of such women between being abused in childhood and suffering depression later in life. Using SPSS or Minitab or solving by hand gives a value for chi-squared of 12.8 with a p-value of 0.002. So we can reject H_0 and conclude that childhood abuse does seem likely to increase the chance of suffering from depression in adult life.

FISHER'S EXACT TEST

- *Used with:* discrete nominal or ordinal data which can take only two values (for example: male or female; above or below the median; agree, disagree; and so on) and when the sample size is too small to use the chi-squared test because of small expected values.
- *Used to test:* similar uses as chi-squared test, i.e. whether there is a relationship between two variables (that is whether or not two variables are independent).

Brief Description of Fisher's Test

Fisher's test calculates the probability of the occurrence of the smallest entry, or one even smaller, in a 2 × 2 contingency table, given that H_0 is true. The procedure for the test is a little too complicated to describe here and interested readers can refer elsewhere for a detailed description[11]. However, it can be used instead of the 2 × 2 chi-squared test:

- whenever the sample size is 20 or less;
- whenever, for samples between 20 and 40, any one expected value is less than 5.

Using a Computer to Do Fisher's Test

Using SPSS to Do Fisher's Test

The procedure with SPSS for Fisher's test is exactly the same as for the chi-squared 2 × 2 test described above. SPSS automatically does Fisher's test as well as chi-squared whenever an expected value is less than 5. I will illustrate its use with an example from practice.

An Example from Practice

Researchers[12] using a randomised clinical trial to investigate the comparative effectiveness of ultrasound (US) versus transcutaneous electrical nerve stimulation (TENS) for painful shoulder syndrome, used Fisher's Exact test (because of the small sample sizes) to check that the demographic characteristics of subjects in the two groups were the same. One of the characteristics checked was the male/female balance in the two groups. The raw data are shown in Table 5.7(a) and the contingency table in Table 5.7(b).

Table 5.7: Data on the male/female balance in two randomised groups for Fisher's Exact test into the comparative effectiveness of ultrasound (US) and transcutaneous electrical nerve stimulation (TENS) in the treatment of painful shoulder syndrome

(a) Raw data

PATIENT	MALE (OR FEMALE)	US (OR TENS)
1	Y	Y
2	Y	Y
3	Y	Y
4	Y	Y
5	Y	Y
6	Y	N
7	N	Y
8	N	Y
9	N	Y
10	N	Y
11	N	Y
12	N	Y
13	N	Y
14	N	Y
15	N	Y
16	N	N
17	N	N
.	.	.
.	.	.
29	N	N

(b) Contingency table

	Male	Female
TENS	1	14
US	5	9

The hypotheses we wish to test are:

H_0: the proportions of males in the US and TENS groups are the same
H_1: the proportions are not the same.

The output is shown in Figure 5.11. Fisher's two-tail test *p*-value is given as 0.08008, compared with the chi-squared value of 0.05365. Neither result is significant if α = 0.05, and we cannot therefore reject H_0. There does not appear to be a significant difference in the proportions of males in each of the two groups. Notice that we also get the message that there are two cells with expected values less than 5.

```
SPSS for MS WINDOWS Release 6.0

US US by MALE Male

                          MALE
                   No            Yes          Row
                    0              1        totals (%)
      US
     No  0         14              1          15
                                             (51.7)
     Yes  1         9              5          14
                                             (48.3)
     Column        23              6          29
   totals (%)    (79.3)         (20.7)      (100.0)

Chi-Square                          Value      DF     Significance
----------                         -------     --     ------------
Pearson                            3.72357      1       0.05365
Continuity Correction              2.16374      1       0.14130
Likelihood Ratio                   3.97223      1       0.04626
Mantel-Haenszel test for           3.59517      1       0.05795
  linear association

Fisher's Exact Test:
One-Tail                                                0.06954
Two-Tail                                                0.08008

Minimum Expected Frequency - 2.897
Cells with Expected Frequency < 5 -   2  of   4 (50.0%)
Number of Missing Observations: 0
```

Figure 5.11 Data on the male/female balance in two randomised groups for Fisher's Exact test into the comparative effectiveness of ultrasound (US) and transcutaneous electrical nerve stimulation (TENS) in the treatment of painful shoulder syndrome

Using Excel to Do Fisher's Exact Test

Fisher's test is not included in the standard version of Excel but is included in the Astute add-in. Excel, unlike SPSS, does not require the original data to be entered but allows the 2 × 2 *contingency table* to be entered directly. The output from Excel gives the values of the cells *a*, *b*, *c* and *d*, the chi-squared statistic, and the one- and two-tailed *p*-values. Space restrictions do not permit me to provide an example.

Using Minitab to Do Fisher's Exact Test

The Minitab package does not calculate Fisher's Exact test.

SUMMARY

This chapter has considered a number of hypothesis tests on nominal data problems for both the one- and two-sample cases. Space does not permit discussion of tests for three or more samples, although for *independent* samples, the chi-squared 2 × 2 test described above may be used in exactly the same way for tables of any size. For *related* or *matched* samples, *Cochran's Q test* may be used. In SPSS this is accessed by clicking on the command sequence:

Statistics
  Nonparametric Tests
    K Related Samples
      ⊙ Cochran's Q

See the *SPSS Base System Users Guide* for further details. Cochran's test is not available in Minitab. In the next chapter we will examine some tests for ordinal variables.

EXERCISES

5.1 Explain how the significance level of a test and the confidence level of a confidence interval estimate are related. What additional information does a confidence interval contain compared with a hypothesis test?

5.2 Discuss the meaning and the nature of the relationship between the significance level of a test, types I and II errors, and the power of a test. How can the power of a test be improved?

5.3 An A&E triage nurse has the impression that on a Saturday night the proportions of males and females arriving in casualty is about the same. He makes a note of the sex of the patients arriving during a half-hour period. These are shown in Table 5.8. Use the binomial test to determine whether the nurse's suspicions are well-founded.

Table 5.8: Sex of 20 consecutive arrivals at an A&E department

F	F	M	M	M	M	M	M	F	F	F	M	F	F	M	F	M	M	M	M

5.4 Table 5.9 refers to a study into the appropriateness of 308 referrals by GPs to a large general hospital, by speciality, judged not to conform to predetermined guidelines[13]. Use the chi-squared goodness-of-fit test to test the hypothesis that the inappropriateness of referrals is uniform across specialities.

5.5 In a study of school children's knowledge of illicit drugs[14], the children were asked "Where did you *first* hear about drugs?" The percentage responses by source of information in 1984 and 1994 are shown in Table 5.10. Use the chi-

Table 5.9: Referrals by GPs to a large general hospital and judged not to conform to guidelines, by speciality

SPECIALITY	REFERRALS	NUMBER JUDGED NOT TO CONFORM TO GUIDELINES
Rheumatology	61	4
Orthopaedics	46	14
Otorhinolaryngology	74	4
Ophthalmology	56	14
Chest medicine	39	4
Gynaecology	32	9

Table 5.10: Sources (%) of information about illicit drugs as recorded by a sample of school children

SOURCE OF INFORMATION	1984 (n = 508)	1994 (n = 384)
Television	82	61
Friends	12	38
Talk in school	8	36
Newspapers	28	25
Parents	12	19
Books	5	10
Radio	6	8

squared goodness-of-fit test to determine whether the percentages observed in each category in 1994 were as might have been expected from the 1984 responses.

5.6 The data in Table 5.11 result from a matched case–control study into bronchial hyper-responsiveness of school children and its possible relationship with exposure to atmospheric sulphur dioxide pollution during infancy. Use McNemar's test to determine whether there is a relationship between pollution and bronchial hyper-responsiveness.

5.7 In a study into the prevalence of HIV infection in patients attending an inner city A&E department[7], the numbers who were HIV positive and those who were also male were recorded. The results in the form of a contingency table are shown in Table 5.12. Use either the chi-squared 2 × 2 test or Fisher's Exact test (whichever is most appropriate) to determine whether there is evidence for a relationship between the two variables.

5.8 In an investigation into the use of a mobile screening unit for the detection of diabetic retinopathy in rural and urban areas[15], the data in Table 5.13 were gathered. Use either a chi-squared 2 × 2 test or Fisher's Exact test (whichever is most appropriate) to determine whether there is any relationship between living in an urban or rural location and symptoms of retinopathy.

5.9 The contingency table shown in Table 5.14 relates to a study[10] into the existence of bulimia in daughters and the existence of childhood sexual and physical abuse. Use a suitable test to determine whether, in these subjects, there is a relationship between the two types of abuse.

Table 5.11: Case–control pairs from investigation into bronchial hyper-responsiveness of school children and relationship with exposure to atmospheric sulphur dioxide pollution during infancy

PAIR	POLLUTED ATMOSPHERE?	
	Cases	*Controls*
1	Y	Y
2	N	N
3	Y	Y
4	Y	N
5	Y	N
6	Y	N
7	N	N
8	N	N
9	Y	N
10	N	Y
11	N	N
12	Y	Y

Table 5.12: Numbers of patients attending an inner-city A&E department[7], the numbers who were HIV positive, and those who were male visitors

MALE	HIV POSITIVE	
	Yes	*No*
Yes	9	451
No	3	455

Table 5.13: Prevalence of diabetic retinopathy in rural and urban areas

EVIDENCE OF RETINOPATHY	RESIDENCE OF SUBJECT	
	Urban (n = 915)	*Rural* (n = 1225)
Yes	384	476
No	531	749

Table 5.14: Data on the existence of bulimia in daughters and the existence of childhood sexual and physical abuse

SEXUAL ABUSE IN CHILDHOOD	PHYSICAL ABUSE IN CHILDHOOD	
	Yes	*No*
Yes	2	3
No	2	4

REFERENCES

1. For example: Daley, L. E. *et al.* (1991) *Interpretation and Uses of Medical Statistics*, Blackwell Scientific; or Kirkwood, B. R. (1988) *Essentials of Medical Statistics*. London: Blackwell Scientific.
2. Altman, D. G. (1991) *Practical Statistics for Medical Research*. London: Chapman & Hall.
3. Silagy, C. *et al.* (1994) Meta-analysis on efficacy of nicotine replacement therapies in smoking cessation. The *Lancet*, **343**, 139–42.
4. Treasure, J. *et al.* (1994) First step in managing bulima nervosa: controlled trial of therapeutic manual. *BMJ*, **308**, 686–9.
5. Nordentoft, M. *et al.* (1993) High mortality by natural and unnatural causes: a 10 year follow up of patients admitted to a poisoning treatment centre after suicide attempt. *BMJ*, **306**, 1637–41.
6. Conway, A. S. *et al.* (1994) The outcomes of targeting community mental health services: evidence from the West Lambeth schizophrenic cohort. *BMJ*, **308**, 627–30.
7. Poznansky, M. C. *et al.* (1994) Prevalence of HIV infection in patients attending an inner city accident and emergency department. *BMJ*, **308**, 636.
8. D'Agostino, R. B. *et al.* (1988) The appropriateness of some common procedures for testing the equality of two independent binomial populations. *The American Statistician*, **42**, 198–201.
9. Green, B. and El-Hihi, M. A. (1990) Out-patient referrals of major depression to psychiatrists in central Liverpool. *Psychiatric Bulletin*, **14**, 465–7.
10. Andrews, B. *et al.* (1995) Depression and eating disorders following abuse in childhood in two generations of women. *British Journal of Clinical Psychology*, **34**, 37–52.7.
11. For example: Daley, L. E. *et al.* (1991) *Interpretation and Uses of Medical Statistics*, London: Blackwell Scientific.
12. Herrera-Lasso, I. *et al.* (1993) Comparative effectiveness of packages of treatment including ultrasound or transcutaneous electrical nerve stimulation in painful shoulder syndrome. *Physiotherapy*, **79**, 251–3.
13. Fertig, A. *et al.* (1993) Understanding variation in rates of referral among general practitioners: are inappropriate referrals important and would guidelines help to reduce rates? *BMJ*, **307**, 1467–70.
14. Denham Wright, J. and Pearl, L. (1995) Knowledge and experience of young people regarding drug misuse, 1969–94. *BMJ*, **310**, 20–4.
15. Leese, G. P. *et al.* (1993) Use of mobile screening unit for diabetic retinopathy in rural and urban areas. *BMJ*, **306**, 187–9.

6

HYPOTHESIS TESTS: ORDINAL VARIABLES

❑ SINGLE-SAMPLE TESTS; THE SIGN TEST; WILCOXON'S SIGNED-RANKS TEST ❑ TWO-SAMPLE TESTS; WILCOXON'S SIGNED-RANKS TEST FOR MATCHED PAIRS; THE MANN–WHITNEY RANK SUM TEST FOR INDEPENDENT PAIRS ❑

YET MORE TESTS

We looked at the main ideas and concepts behind hypothesis testing in the previous chapter, and examined a number of tests which can be used with nominal data. In this chapter I will describe three of the most widely used tests which can be used when the data are at least at the ordinal level. Most of these tests are concerned with hypotheses about the value of the population median and are applied to ordinal data which has first been ranked. I will start with two tests for a single sample or group.

SINGLE-SAMPLE TESTS

The Sign Test

- *Used with:* data which are *at least* ordinal, but which cannot be quantified other than that they are above or below some value, e.g. above or below a hypothesised median value. The sign test is identical to the binomial test which we looked at in Chapter 5, except that in the sign test the data are expressed in frequencies, in the binomial test they are given in proportions.
- *Used to test:* that the median of some single population is equal to some hypothesised value.

Brief Description of the Sign Test

Our reluctance or inability to quantify the data may be due, for example, to our unhappiness with the way in which a scale has been designed or interpreted by the subject, or in its suitability in any particular example, or by the fact that quantitative measurements are simply not available.

It is easiest to describe the sign test using an example. Suppose we ask a group of ex-patients, recently discharged from hospital after investigations for chest pain, to assess the relative contribution of the nurses and the doctors to their treatment and care[1]. A 10 cm visual analogue scale is used with the left-hand end marked "Entirely due to nurses" and the right-hand end marked "Entirely due to doctors". The scale is subsequently overlaid with a rule scaled from 0 to 10 and each patient's score thus determined. The scores are clearly ordinal. Our null hypothesis is that the population of such ex-patients believe that nurses and doctors contribute equally to their treatment and care, i.e. the null hypothesis H_0 is that the population median score is 5. The alternative hypothesis H_1 is that the median score is not 5. In other words:

H_0: Median = 5
H_1: Median ≠ 5.

If the null hypothesis is correct we should find as many scores below the median score of 5 as above it. If we mark each score above the median as "+" and each score below the median as "–" then there should be as many pluses as there are minuses. In other words the probability of getting a plus is the same as the probability of getting a minus. If there are too few (or too many) pluses or minuses, then we may reject H_0. As it happens the probability of getting some given number of pluses in a sample of given size is given by the binomial distribution, described previously. The sign test examines the number of values above and below the hypothesised median (ignoring those equal to this value) and uses binomial probabilities to decide whether to accept or reject the null hypothesis. The sign test is more often used in a two-sample situation as we will discover shortly, but it can be very useful in single-sample examples such as this.

Using a Computer to Do the Sign Test

Let us assume that in the above example we interview a sample of 32 ex-patients and obtain the values shown in column 1 of Table 6.1. Column 2 shows whether the value is above (+) or below (–) the hypothesised median of 5.

Table 6.1: Scores recorded by a sample of 32 ex-patients using a 10-point visual analogue scale: the null hypothesis is that the median score is equal to 5

PATIENT	SCORE	DIRECTION	PATIENT	SCORE	DIRECTION
1	6	+	17	2	–
2	10	+	18	5	=
3	2	–	19	6	+
4	1	–	20	5	=
5	2	–	21	1	–
6	3	–	22	6	+
7	6	+	23	5	+
8	1	–	24	5	+
9	0	–	25	1	–
10	5	=	26	1	–
11	6	+	27	10	+
12	3	–	28	2	–
13	3	–	29	2	–
14	5	=	30	5	=
15	0	–	31	5	=
16	5	=	32	1	–

So there are 16 minuses and 9 pluses, implying that 16 patients thought nursing staff, and nine thought doctors, respectively, contributed most to their care and treatment, with seven ex-patients who thought that both nurses and doctors played an equal role, i.e. there were seven scores equal to the hypothesised median value of 5. These equal scores are discarded and *not* included in the test. If the null hypothesis is true we would have expected that the number of pluses and minuses would have been the same. In this case they are not, but is this departure from equal numbers large enough to cause us to reject H_0, or could it have happened by chance alone?

Using Minitab to Do the Sign Test

Minitab doesn't allow us to enter pluses and minuses into the worksheet. Instead, we have to enter 16 values less than the null hypothesis value of 5, nine greater than 5, and seven values of 5. The actual above and below values entered are completely arbitrary in that they make no difference to the *p*-value. I entered 16 values of 4, and 9 values of 6 in column 1 and named it "VASscore". The following commands will produce the output shown in Figure 6.1:

Stat
　Nonparametrics
　　1-Sample Sign
　　　Select c1
　　　⊙ Test median
　　　　Change default H_0 from 0.0 to 5.0
　　　　　(Accept: Alternative "not equal to")
　　　　　　OK

```
MTB > STest 5.0 ''VASscore'';
SUBC> Alternative 0.

Sign Test for Median

Sign test of median = 5.000 versus N.E. 5.000

              N    Below    Equal    Above    P-value   Median
VASscore     32       16        7        9     0.2295    4.500
```

Figure 6.1: Output from Minitab for the one-sample sign test of the belief of patients about the relative contributions of doctors and nurses to their hospital care and treatment

The Minitab output records the number of values above, equal to, and below the hypothesised median and calculates the sample median value (this will of course depend on the values we enter, and so must be ignored, since these are arbitrary). If we set a significance level $\alpha = 0.05$, then the *p*-value of 0.2295 is clearly not significant and we thus accept the null hypothesis that the median score is 5, i.e. that ex-patients believe nurses and doctors contribute equally to the care and treatment. The *p*-value here is the probability of getting nine or fewer, or 14 or more, plus values out of 25.

SPSS cannot be used to perform the *one*-sample sign test.

The Wilcoxon Signed-ranks Test

- *Used with:* data which are *at least* ordinal.
- *Used to test:* that the median of a single population is equal to some hypothesised value.

Brief Description of the Wilcoxon Signed-ranks Test

Whereas the sign test ignored the actual data values and simply measured their *direction,* i.e. whether they were above or below some hypothesised value, Wilcoxon's signed-ranks test takes not only the direction of the sample values into account but also their *magnitude*. The Wilcoxon test is a powerful alternative to the *t* test for metric data (which I will describe in the next chapter) if the distribution is known not to be Normal, or when the sample size is too small to determine whether it is Normal or not.

Like many ordinally based tests, the Wilcoxon test is applied to ranked values. The hypothesised median is subtracted from each sample value, values equal to the hypothesised median are ignored and *all* of the remaining differences are then ranked (ignoring any + or – signs). A rank of 1 is given to the smallest value, a rank of 2 to the next smallest, and so on (values which are the same are given their average rank). Then all of the ranks which came from negative differences are summed, and all of the ranks which came from positive differences are summed. If the null hypothesis value for the median is correct we should expect to find that the sum of the negative ranks is about the same as the sum of the positive ranks, since the negative and positive scores should be about the same in number and size.

If we apply this procedure, using a hypothesised median of 5, to the data in the example above on the views of the ex-patients about their care and treatment, in Table 6.1, we will get the values shown in Table 6.2. Ranks which come from negative differences are in brackets, ().

Table 6.2: Ranked VAS scores of patients' views on the relative contributions of nurses and doctors to their care and treatment while in hospital for chest pain

Patient	Score	Difference from 5*	Rank	Patient	Score	Difference from 5*	Rank
1	6	1	3	17	2	(3)	(12)
2	10	5	23.5	18	5	=	
3	2	(3)	(12)	19	6	1	3
4	1	(4)	(18.5)	20	5	–	
5	2	(3)	(12)	21	1	(4)	(18.5)
6	3	(2)	(7)	22	6	1	3
7	6	1	3	23	5	=	
8	1	(4)	(18.5)	24	5	=	
9	0	(5)	(23.5)	25	1	(4)	(18.5)
10	5	=		26	1	(4)	(18.5)
11	6	1	3	27	10	5	23.5
12	3	(2)	(7)	28	2	(3)	(12)
13	3	(2)	(7)	29	2	(3)	(12)
14	5	=		30	5	=	
15	0	(5)	(23.5)	31	8	3	12
16	8	3	12	32	1	(4)	(18.5)

* () are–ve

Sum of positive ranks: 3 + 3 + 3 + 3 + 3 + 12 + 12 + 23.5 + 23.5 = 86

Sum of negative ranks: 7 + 7 + 7 + 12 + 12 + 12 + 12 + 12 + 18.5 + 18.5 + 18.5 + 18.5 + 18.5 + 18.5 + 23.5 + 23.5 = 239

So the sum of the positive ranks is 86 and the sum of the negative ranks is 239*. If the null hypothesis, that the median value is 5, is true, then we would expect these sums to be about the same. Clearly they are not the same and the crucial question is: are they close enough for us to accept H_0? To do the test by hand, we denote the smallest of the two rank sums as T and then refer to Table A6 in the Appendix. For the test to be significant, the value of T must be *equal to or less than* the value found in Table A6, in the column corresponding to the chosen level of significance, and in the row equal to n minus the number of ties (i.e. n minus the number of scores equal to the hypothesised value), which is $32 - 7 = 25$.

In this example, $T = 86$ and the values in row 25 are 90 (for $\alpha = 0.05$) and 68 (for $\alpha = 0.01$). Since 86 is less than 90 (but not less than 68) the test is significant at $\alpha = 0.05$ but not significant at $\alpha = 0.01$. Thus, if our chosen level of significance is 0.05 we can reject the null hypothesis that the median score of ex-patients is 5, i.e. that nurses and doctors contribute to care and treatment equally. The sample evidence, 16 out of 25 ex-patients judging nurses to have contributed more to their care and treatment than doctors, compared with only nine favouring the doctors' role, might suggest that it is nurses who on the whole play a more important role, although the alternative hypothesis is simply that their contribution is not the same. The result we obtained using Wilcoxon's test is different from the result of the sign test above. The explanation for this is that the Wilcoxon test uses more of the information in the sample data, taking into account the size as well as the direction of the differences of each score from the hypothesised median.

Using a Computer to Do the Wilcoxon One-sample Signed-ranks Test

Using Minitab to Do the Wilcoxon One-sample Signed-ranks Test

To do Wilcoxon's one-sample signed-ranks test with Minitab we first enter the scores into column c1, say, of the worksheet and then use the commands:

Stat
  Nonparametrics
    1-Sample Wilcoxon
      Select c1
        ⊙ Test median
          Change default H_0 from 0.0 to 5.0
            (Accept: Alternative "not equal to")
              OK

This will produce the output shown in Figure 6.2. The resulting value for T is 86, the same as the by-hand calculation above, and the p-value is calculated to be 0.041, less than 0.05, and thus enabling us to reject H_0. The sample median is calculated to be 3.5, indicating perhaps a more important contribution by nurses. The p-value here is the probability of getting as small, or smaller, sum of positive ranks (or negative ranks, whichever is the smaller value) as 86 out of a total sum of positive and negative ranks of 325.

* As a useful check if you have to do this procedure by hand, the sum of the positive and negative ranks is $n(n + 1)/2$, which in this example is $(25 \times 26)/2 = 325$, which is the same as $86 + 239$.

```
MTB > WTest 5.0 ''VASscore'';
SUBC> Alternative 0.

Wilcoxon Signed Rank Test

TEST OF MEDIAN = 5.000 VERSUS MEDIAN N.E. 5.000

                    N for    Wilcoxon                Estimated
              N     test     statistic    P-value     median
VASscore     32      25        86.0        0.041       3.500
```

Figure 6.2: Output from Minitab for the one-sample Wilcoxon signed-ranks test of the belief of patients about the relative contributions of doctors and nurses to their hospital care and treatment

Using SPSS to Do the Wilcoxon One-sample Test

SPSS cannot do Wilcoxon's signed-ranks test on a single sample, since it is necessary to select *two* variables in the dialogue box for this test.

TESTS WITH TWO MATCHED SAMPLES

The Two-sample Sign Test

- *Used with:* two matched or paired samples or groups where the data in each sample or group are *at least* ordinal. The sign test tends to be used where, although it may be difficult actually to quantify the scores, it is possible to say which score in each pair is the greater.
- *Used to test:* whether the population medians of two groups are equal, i.e. the difference in the medians is zero. The null hypothesis is H_0: median 1 – median 2 = 0.

Brief Description of the Two-sample Sign Test

The sign test I described at the beginning of this chapter for a single sample is in practice more often used as here with two matched samples. Nonetheless, the principle of the two-sample sign test is exactly the same as the one-sample test described above. The test requires that we must be able to say which score in each pair of scores is the greater. If the score from the first group is greater this pair receives a plus sign, if the score from the second group is greater the pair receives a minus sign. If the null hypothesis, that the difference between the medians of the two groups is zero, is true, then we expect that the number of pairs where the first value is greater should be the same as the number of pairs where the second value is greater. In other words, for each pair the probability of getting a plus is the same as the probability of getting a minus, and as we have seen we can use the binomial distribution to determine the probability (i.e. the p-value) of getting any number of plus signs in a given sample (excluding ties), and hence test the null hypothesis.

Using a Computer to Do the Two-sample Sign Test

Using SPSS to Do the Two-sample Sign Test

I am going to illustrate the use of a computer to do the sign test with some real data on the Barthel Activities of Daily Living (ADL) scores for a sample of 30 hospitalised geriatric patients*, collected routinely each week before and after remedial treatment. Before-and-after groups are a common example of matched data. The data for a week chosen at random are shown in Table 6.3. Note that there are two pairs where the before-and-after scores are the same; these tied scores are ignored in the test.

Table 6.3: Before-and-after Barthel scores for a sample of 30 hospitalised geriatric patients

PATIENT	BEFORE	AFTER	PATIENT	BEFORE	AFTER
1	11	19	16	8	5
2	10	16	17	7	7
3	0	17	18	14	20
4	4	3	19	14	18
5	6	16	20	12	13
6	8	11	21	12	12
7	13	15	22	16	12
8	6	15	23	20	19
9	6	10	24	8	4
10	14	8	25	9	13
11	12	19	26	10	17
12	16	17	27	12	9
13	1	10	28	8	12
14	8	16	29	10	17
15	13	17	30	2	11

The null hypothesis is that the median Barthel score before remedial treatment is the same as that after, i.e. there is no difference between them. That is:

H_0: median Barthel before – median Barthel after = 0
H_1: median Barthel before – median Barthel after $\neq$ 0.

The data are entered into columns 1 and 2, say, of the SPSS datasheet and named "before" and "after" using **Data, Define Variables**. The command sequence is then:

Statistics
  Nonparametric Tests
    2 Related Samples
      Select before and after (in the Test Pair(s) List box)
        Wilcoxon button (to remove) in Test Type box
          ☒ **Sign button** (to include) in Test Type box
          OK

---

* The Barthel Activities of Daily Living (ADL) Scale is a measure of the degree to which an individual has the ability to live independently. It measures a subject's performance on 10 factors, including: mobility, climbing stairs, bathing, incontinence, dressing, feeding, etc. The range of scores is from 0 (completely dependent) through 10 (moderately dependent) to 20 (completely independent).

The output produced is shown in Figure 6.3. SPSS uses the Normal approximation to the binomial for sample sizes over 25. The calculated p-value of 0.0376, the probability of getting as few as eight (or fewer) plus signs, out of 28 values, allows us to *reject* the null hypothesis that there is no difference between the before-and-after Barthel scores in populations for whom these samples are representative. The remedial treatment does appear to increase significantly the ADL score of such patients.

```
SPSS for MS WINDOWS Release 6.0

Sign Test

AFTER with BEFORE

   Cases
    20      - Diffs (BEFORE LT AFTER)      Z =   2.0788
     8      + Diffs (BEFORE GT AFTER)
     2      Ties                            2-Tailed
                                             P = .0376
    __
    30    Total
```

Figure 6.3: Output from SPSS for the two-sample sign test on before-and-after Barthel scores for a sample of 30 geriatric patients

Using Minitab to Do the Two-sample Sign Test

The use of Minitab for the two-sample sign test is a little more complicated than with SPSS. Minitab only has the facility for a *one*-sample sign test. This means that the differences between the before and after scores have to be calculated first before the test is used (unlike SPSS which does the necessary calculation as part of its sign test procedure). The before-and-after scores are first entered into columns 1 and 2, say, of the worksheet. The subtraction of the before scores from the after scores can then be achieved either by using the **let** command in the sessions window (as described previously), or with the commands:

Calc
  Mathematical Expressions

This brings up the mathematical dialogue box. Type c3 in the **Variable (new or modified)** box, then type **after – before** in the **Expression:** box. Then click on **OK**. This will enter the differences in column c3, which can then be named "Diff".

To perform the sign test on the values in the "Diff" column the following commands are used:

Stat
  Nonparametrics
    1-Sample Sign
      Select "Diff"
        ⊙ Test median button
          (Accept default H_0 value of 0.0)
            (Accept default Alternative of "not equal to")
              OK

This will produce the output shown in Figure 6.4. Minitab uses the more exact binomial distribution for sample sizes up to 50. This explains the difference in the Minitab *p*-value of 0.0125 compared with SPSS's Normal approximation *p*-value of 0.0376. So the test is still significant if $\alpha = 0.05$.

```
MTB > Let c4 = ''After''-''Before''

MTB > STest 0.0 ''Diff'';
SUBC> Alternative 0.

Sign Test for Median

Sign test of median = 0.00000 versus N.E. 0.00000

             N    Below    Equal    Above    P-value   Median
Diff        30        7        2       21     0.0125    4.000
```

Figure 6.4: Output from Minitab for the sign test for the difference in the median before-and-after Barthel scores

The Two-sample Wilcoxon Matched-pairs Signed-ranks Test

The basic features of this test are identical to those described in the discussion of the one-sample Wilcoxon test above. Like the two-sample sign test it is used to test the null hypothesis that there is no difference in the medians of two matched populations (i.e. the medians are the same), but the Wilcoxon test uses not only the direction of the difference in the two sets of data but also the *size* of those differences.

The difference in each pair of scores is calculated, and these differences are ranked (ignoring signs) ranking the smallest difference as 1, the next smallest difference as 2, and so on. The sums of the ranks deriving from positive differences and from negative differences are determined separately, and the smallest of these sums, called *T*, is compared with the table of values in Table A6.

If we apply this procedure to the data in Table 6.3, which we used for the sign test, we get Table 6.4. From this the sum of positive ranks is 346 and the sum of negative ranks is 60. Thus $T = 60$. There are 30 observations in the sample and two ties so the row is $30 - 2 = 28$. The values in row 28 of Table A6 are 117 (0.05) and 92 (0.01). Since 60 is less than 92 we can say that the test is significant at better than 0.01 and reject H_0. There does appear to be a significant difference in the median before-and-after Barthel scores.

An Example from Practice

In a study[2] to evaluate the efficacy of a pilot service offering therapy specifically to adults with a history of child sexual abuse, 59 subjects completed three psychological questionnaires before and after receiving therapy. These were: the social activities and distress scale (SADS); the general health questionnaire (GHQ); and the delusions, symptoms and

Table 6.4: Calculation of T for the Wilcoxon matched-pairs signed-ranks test on the before-and-after Barthel scores for a sample of 30 hospitalised geriatric patients

PATIENT	BARTHEL SCORE		DIFFERENCE	RANKED DIFFERENCES
	Before	*After*	*After – Before*	() are–ve
1	11	19	8	22.5
2	10	16	6	17
3	0	17	17	28
4	4	3	–1	(2.5)
5	6	16	10	27
6	8	11	3	7
7	13	15	2	5
8	6	15	9	25
9	6	10	4	12
10	14	8	–6	(17)
11	12	19	7	20
12	16	17	1	2.5
13	1	10	9	25
14	8	16	8	22.5
15	13	17	4	12
16	8	5	–3	(7)
17	7	7	0	–
18	14	20	6	17
19	14	18	4	12
20	12	13	1	2.5
21	12	12	0	–
22	16	12	–4	(12)
23	20	19	–1	(2.5)
24	8	4	–4	(12)
25	9	13	4	12
26	10	17	7	20
27	12	9	–3	(7)
28	8	12	4	12
29	10	17	7	20
30	2	11	9	25

Sum of positive ranks = 346; sum of negative ranks = 60; $T = 60$

states inventory (DSSI). Wilcoxon's matched-pairs signed-ranks test (two-tailed) was used to test the significance of the difference between the scores on each scale between the start and end of therapy. The null hypothesis was that there was no difference between the before-and-after scores. Wilcoxon's test is clearly appropriate here because the data are ordinal and the two samples are self-matching. Notice that the authors followed good practice in using a two-tailed test even though they had reason to believe that there was an *improvement* in the scores.

The authors reported significant differences in all three measures. The *p*-values were as follows: in the SADS scores, $p = 0.001$; in the GHQ, $p < 0.00001$; in the DSSI scores, $p < 0.00001$.

Using a Computer to Do the Two-sample Wilcoxon Test

We can illustrate the use of a computer to do Wilcoxon's test using the before-and-after Barthel scores from Table 6.3.

Using SPSS to Do the Two-sample Wilcoxon Test

The commands in SPSS for the Wilcoxon test are the same as for the two-sample sign test except that we accept the default Wilcoxon test in the Test Type box:

Statistics
  Nonparametric Tests
    2 Related Samples
      Select before and after (in the Test Pair(s) List box)
        ☒ **Wilcoxon**
          OK (accepts default test and Alternative "not equal to")

These commands produce the output shown in Figure 6.5. The *p*-value is calculated to be 0.0011, so the test is significant (0.0011 is less than $\alpha = 0.01$) and we can reject H_0. The same decision we reached by hand above, although now we have a much more precise *p*-value.

```
SPSS for MS WINDOWS Release 6.0

Wilcoxon Matched-Pairs Signed-Ranks Test

AFTER with BEFORE

   Mean Rank      Cases
     16.48         21 -      Ranks (BEFORE LT AFTER)
      8.57          7 +      Ranks (BEFORE GT AFTER)
                    2        Ties  (BEFORE EQ AFTER)
                   --
                   30      Total
                    Z =    -3.2563     2-Tailed P = 0.0011
```

Figure 6.5: Output from SPSS for the two-sample Wilcoxon signed-ranks test for the difference in the median before-and-after Barthel scores

Using Minitab to Do the Two-sample Wilcoxon Test

We face the same problem with Minitab as we did with its two-sample sign test procedure. Minitab only does a one-sample Wilcoxon test and so we have first to calculate a column of "before – after" difference values before we can apply the Wilcoxon test to this column of difference values. If we use:

Calc
  Mathematical Expressions

to do this, as we did with the sign test above, naming the column of differences "diff", and then follow the command sequence:

Stat
  Nonparametrics
    1-Sample Wilcoxon
      Select "Diff"
        ⊙ **Test median**
          Accept default H_0 value of 0.0

Accept default Alternative of "not equal to"
OK

we obtain the output shown in Figure 6.6. The p-value is 0.001, and Minitab also calculates the sample median value of 3.5 for the differences.

```
MTB > WTest 0.0 ''Diff'';
SUBC> Alternative 0.

Wilcoxon Signed Rank Test

TEST OF MEDIAN = 0.000000 VERSUS MEDIAN N.E. 0.000000

                    N for     Wilcoxon                Estimated
            N       test      statistic     P-value     median
Diff        30      28        346.0         0.001       3.500
```

Figure 6.6: Output from Minitab for the Wilcoxon signed-ranks test of the difference in the median before-and-after Barthel scores

Using Excel to Do the Two-sample Wilcoxon Test

The standard version of Excel does not include the Wilcoxon signed-rank test, but the Astute add-in does allow it. We can illustrate its use with the Barthel data. If the two columns of data are entered into columns A and B, selected, and the Wilcoxon test button on the Astute toolbar clicked on, the output shown in Figure 6.7 is produced.

```
                   n      Rank Sum       Mean Rank

Positive:          7        60             8.571

Negative:         21       346            16.476

Excluded:          2

Wilcoxon R:       60

z:    -3.4894
p:    0.0005 corrected for ties
```

Figure 6.7: Output from the Extel Astute add-in for the Wilcoxon signed-ranks test on the Barthel data

Astute corrects for ties and the p-value of 0.0005 differs from the SPSS (0.0011) and Minitab (0.001) values accordingly. In any case the same decision to reject the null hypothesis of no relation between the variables is reached.

TESTS WITH TWO INDEPENDENT SAMPLES

The Mann–Whitney Rank Sum Test

- *Used with:* data which are *at least* ordinal.
- *Used to test:* the Mann–Whitney test is used to test whether two independent distributions are identical, so it can be used to test for either means or medians. In most applications, however, it is used as a test of whether two population *medians* are equal.

Brief Description of the Mann–Whitney Test

The Wilcoxon signed-ranks test described above is one of the most useful and widely used tests for examining the difference in the medians of two matched populations when the data are either ordinal, or are metric but not Normally distributed (i.e. are non-parametric). The Mann–Whitney U rank sum test* is equally important for examining the difference in the medians of two *independent* populations when the data are either ordinal or metric and not known to be Normal.

I will illustrate the Mann–Whitney test procedure using the data in Table 6.5 on the age at which male and female students at Oxford University committed suicide between 1976 and 1990[3]. These two samples are independent, and the data are metric but too small to determine whether suicide age in either male or female students is Normally distributed. For these reasons the Mann–Whitney non-parametric test seems appropriate. The null hypothesis is that the median age at which male and female students commit suicide is the same, i.e. there is no difference between their median ages:

H_0: Median male age – median female age = 0
H_1: Median male age – median female age ≠ 0.

Table 6.5: The ages at which male and female students at Oxford University committed suicide between 1976 and 1990

MALE AGES	MALE RANKS R_M	FEMALE AGES	FEMALE RANKS R_F
18	1	21	7.5
19	2	22	13
20	3.5	23	17.5
20	3.5	23	17.5
21	7.5	24	20
21	7.5		
21	7.5		
21	7.5		
21	7.5		
22	13		
22	13		
22	13		
22	13		
23	17.5		
23	17.5		
25	21		

Sum of male ranks $\Sigma R_M = 155.5$; sum of female ranks $\Sigma R_F = 75.5$

* Also developed independently by Wilcoxon whose approach is known as the *Wilcoxon ranked-sum* W test.

I understand it's called the Mann–Whitney–Houston test.

The first step is to combine the 16 male and 5 female scores (the ages) and rank them as if they were *one sample*, giving the smallest value rank 1, the second smallest rank 2, and so on. The group from which each rank came should be noted. Unlike with the Wilcoxon test the sign of each value is taken into account. For example, –2 would rank below –1, which would rank below 1, etc. As it happens, there are no negative values in this example. The ranks for the male and female students are shown in the table, and are then separately summed to produce the rank sums, $\Sigma R_M = 155.5$ and $\Sigma R_F = 75.5$.

To perform the Mann–Whitney test we make use of what is called the Mann–Whitney U statistic. We'll come to the meaning of U in a moment but first, if the sample sizes of the male and female students are n_M and n_F respectively, then U can be calculated from *either* one of the following rather intimidating expressions:

$$U = n_M n_F + \frac{n_M(n_M + 1)}{2} - \Sigma R_M$$

or

$$U = n_M n_F + \frac{n_F(n_F + 1)}{2} - \Sigma R_F.$$

These may look a little off-putting at first sight but, if a suitable computer program which can do the Mann–Whitney test is not available, there is no avoiding their use. Let's first put the appropriate numeric values into the first equation, $n_M = 16$, $n_F = 5$ and $\Sigma R_M = 155.5$. The expression then becomes:

$$U = 16 \times 5 + \frac{16 \times (16 + 1)}{2} - 155.5.$$

The answer is $U = 60.5$. If we repeat this procedure for the second equation above we find $U = 19.5$. We use the *smallest* U value to perform the hypothesis test, i.e. $U = 19.5$.

The meaning of U is a bit more difficult[4], but basically it is the number of pairs of sample values where the value from the first sample is smaller than the value from the second sample. Without getting any more technical, U can then be used to determine the p-value, i.e. the probability of getting the sum of ranks we do get (or one more extreme) given that the null hypothesis (no difference in medians) is true.

A table of critical values is given in Table A7 in the Appendix at the end of this book, for sample sizes up to 25, for $\alpha = 0.05$ and $\alpha = 0.01$. U must be *equal to or smaller* than the value in Table A7 to be significant, i.e. for us to reject H_0. In this example, with a smaller sample size of 5 and a larger sample size of 16, we find in Table A7 values of 15 for $\alpha = 0.05$ and 9 for $\alpha = 0.01$. Since the U value of 19.5 is greater than both of these values, we cannot reject H_0 and we would accept that there is no difference in the median ages at which such male and female students commit suicide.

An Example from Practice

In a study[5] to compare the use of patient controlled analgesia (PCA) with traditional intramuscular injection (IM) for the management of postoperative pain relief among patients who had undergone major abdominal surgery, the researchers used PCA on one ward (28 patients) and IM on a second ward (21 patients). No attempt was made to match patients in any way. Patients were asked to indicate their degree of pain 72 hours after the operation using a visual analogue scale, a 100 mm straight line marked at one end with "no pain" (0) and at the other with "very severe pain" (100).

The researchers used the Mann–Whitney test to compare the levels of pain on the two wards. Mann–Whitney is appropriate because the data are ordinal (VAS scale) and the two samples (ward 1 and ward 2) are independent. The null hypothesis is that there is no difference in median pain levels achieved using PCA and IM. The test found that the difference in the two median pain scores was not significant, $U = 249$, $p > 0.05$, and so the null hypothesis could not be rejected.

Using a Computer to Do the Mann–Whitney test

Using SPSS to Do the Mann–Whitney Test

We can use the student suicide data in Table 6.5 as an illustration. SPSS assumes that all of the data are in one column, so male followed by female ages are entered into column 1, say. Column 2 contains a grouping variable, 1 in each of the first 16 rows (= male), and 2 in each of the next five rows (= female). The command sequence is then:

Statistics
Nonparametric Test
2-Independent Samples
Select c1 for Test Variable List
Select c2 for Grouping Variable
☒ **Mann-Whitney U**
Define Groups
type **1** and **2** (in Define Groups dialogue box)
OK (accepts default)

The output produced is shown in Figure 6.8. SPSS calculates the same *U* value of 19.5 that we just obtained by hand. The Mann–Whitney procedure described above is fairly sensitive to the number of tied values (of which there are quite a few in this example), and SPSS provides two *p*-values, one uncorrected for ties, 0.0910, and one corrected for ties, 0.0836. As it happens, both of these values exceed 0.05 which confirms the result obtained above, that we cannot reject H_0 if $\alpha = 0.05$. There does not appear to be any significant difference in the median ages at which such male and female students commit suicide.

```
SPSS for MS WINDOWS Release 6.0

Mann–Whitney U – Wilcoxon Rank Sum W Test

AGE by SEX

  Mean Rank    Cases
     9.72       16          SEX = 1.00 male
    15.10        5          SEX = 2.00 female
                --
                21    Total

                          Exact          Corrected for ties
     U         W       2-Tailed P          Z           2-Tailed P
   19.5      75.5        .0910          -1.7302          0.0836
```

Figure 6.8: Output from SPSS for the Mann–Whitney test on the difference in median ages of male and female students committing suicide

Using Minitab to Do the Mann–Whitney Test

Minitab uses the equivalent Wilcoxon rank sum test and calculates Wilcoxon's *W* statistic rather than the Mann–Whitney *U*. Minitab assumes that the data for the two groups are in two separate columns, so I entered male ages in column c1 and female ages in column c2. The command sequence for the Mann–Whitney test is then:

Stat
Nonparametrics
Mann-Whitney
Select c1 (as First Sample)
Select c2 (as Second Sample)
OK (accepts default Alternative of "not equal to")

The output produced by Minitab is shown in Figure 6.9. Minitab first calculates a 95.7% confidence interval for the difference between the two medians, which it labels ETA1 and ETA2 (ETA, written η, is a letter of the Greek alphabet), of (–3.00 to –0.00). (The confidence interval calculation requires that the data be metric.) This interval includes zero, which implies no difference between the two medians. This is confirmed by the results of the hypothesis test which shows p-values of 0.0986 (not corrected for ties) and 0.0914 (corrected for ties). These are very similar to the SPSS values above, and, since both values are greater than 0.05, do not allow us to reject H_0. Minitab does not provide a value for U although it does provide a value for the larger sum of ranks as $W = 155.5$. The slight difference in the p-values between SPSS and Minitab may be accounted for by the slightly different approaches, rounding errors, approximations, etc.

```
MTB > Mann–Whitney 95.0 ''Males'' ''Females'';
SUBC> Alternative 0.

Mann–Whitney Confidence Interval and Test

Males     N = 16     Median = 21.000
Females   N =  5     Median = 23.000

Point estimate for ETA1–ETA2 is –1.000

95.7 Percent C.I. for ETA1–ETA2 is (–3.000,–0.000)

W = 155.5

Test of ETA1 = ETA2 vs. ETA1 ~= ETA2 is significant at 0.0986

The test is significant at 0.0914 (adjusted for ties)

Cannot reject at alpha = 0.05
```

Figure 6.9: Output from Minitab for the Mann–Whitney U (Wilcoxon W) test on the difference in median ages of male and female students committing suicide

Using Excel to Do the Mann–Whitney Test

The standard version of Excel does not include the Mann–Whitney test but if the Astute add-in is available that can be used, and we can apply the test to the student suicide data. If these data are entered in columns A and B of the spreadsheet, and the Mann–Whitney button on the Astute toolbar clicked on, the output in Figure 6.10 is produced.

The p-value produced is exactly the same (0.0836) as that from SPSS and Minitab, confirming non-rejection of the hypothesis that the mean age of the male and female students is the same.

USING THE CHI-SQUARED TEST WITH ORDINAL DATA

Before we end this chapter we should note that the chi-squared test can be used to test the existence of a relationship (which it does you will recall from Chapter 5,

```
          n      Rank Sum    Mean Rank      U
  A:     16       155.5        9.719       60.5
  B:      5        75.5       15.100       19.5

Mann—Whitney 19.5

z: 1.7302
p: 0.0836 corrected for ties
```

Figure 6.10: Output from the Extel Astute add-in for the Mann–Whitney test applied to the student suicides data

by comparing the proportions in each category of two separate groups) between two independent groups or samples when the categories are at the ordinal level. The procedure is exactly as described in Chapter 5, where we examined a number of examples using nominal categories. As an example, the contingency table in Table 6.6 is taken from a study into the beliefs about what health promotion education involves among traditional and Project 2000 student nurses[6].

Table 6.6: Beliefs about what health promotion education involves in a study among student nurses: responses to the question "Should health promotion education be about attempting to modify unhealthy behaviour?"

STUDENTS	STRONGLY AGREE	AGREE	NOT SURE	DISAGREE	STRONGLY DISAGREE
Traditional	8	34	0	1	0
Project 2000	13	23	6	0	1

If you were doing this chi-squared test by hand you would need to recall that the number of degrees of freedom (the row we have to use when we look up the critical value in the chi-squared table) is equal to:

(no. rows – 1) × (no. columns – 1).

The number of d.f. in this 2 × 5 table is therefore 1 × 4 = 4. So we would use row 4. The authors report a value for $\sum \frac{(O-E)^2}{E}$ of 4.96, which they state is "significant at the 5% level" but do not unfortunately provide the exact p-value (which I have left as an exercise). The null hypothesis (not explicitly stated by the authors) is that there is no relationship between the two variables, namely, view of health promotion, and type of student nurse. In other words, the views of the traditional and Project 2000 nurses are identical because the proportions in each category among the two types of nurses is the same. It appears that this hypothesis can be rejected if $\alpha = 0.05$.

I should mention that the chi-squared test can also be used as a test of trend. In this example it would be a test of whether the views of the two groups of nurses got further apart (or closer together) as we moved from "Strongly agree" through to "Strongly disagree". Unfortunately there is not the space to consider this, but it is described in many other sources[7].

SUMMARY

In this chapter we have examined a number of tests which can be used if the data are at least at the ordinal level. This discussion has included two of the most important tests used to discriminate between two population medians, the Wilcoxon rank sum test for matched pairs, and the Mann–Whitney signed-ranks test for independent pairs. Both of these tests are non-parametric, i.e. they make no assumptions about the shape of the distribution of the variables involved. In the next chapter I am going to discuss the *t* test for both one sample and two samples (matched and independent). This is a parametric test that *does* require that the variables in question be Normally distributed.

EXERCISES

6.1 In a study into the satiating effect of meals supplemented with either fat or carbohydrate[8], the investigators used as subjects 16 lean healthy male volunteers who were extensively screened before entry to the study to exclude any subject with a history of eating disorders, or a tendency to restrict food intake. Dietary restraint was assessed by the three-factor Eating Inventory Test (EIT), and disordered eating by the Eating Attitudes Test (EAT). The baseline scores of the 16 subjects included in the study on both of these tests is given in Table 6.7. Use (a) the one-sample sign test, and (b) the one-sample Wilcoxon test, to test the hypothesis that the median score on each test is zero.

Table 6.7: Baseline EAT and EIT scores of 16 subjects in a study into the satiating effect of meals supplemented with either fat or carbohydrate

SUBJECT	EAT SCORE	EIT SCORE	SUBJECT	EAT SCORE	EIT SCORE
1	0	1	9	4	6
2	4	1	10	0	3
3	3	6	11	2	3
4	0	3	12	1	2
5	0	1	13	1	6
6	0	1	14	1	1
7	1	1	15	5	7
8	0	0	16	7	8

6.2 In a study into the effects of activity nursing on the quality of life of elderly patients in a continuing care unit[9], the cognitive functioning of the patients was measured before, and eight months after, the introduction of activity nursing, using a 13-question modified mental test score. The results are shown in Table 6.8. Use both the two-sample sign and Wilcoxon's tests to determine whether there was a significant improvement in cognitive function. How did the change in the female patients compare with that in the men?

6.3 In a study into the dietary requirements of children suffering from cerebral palsy[10], the energy intake of two small independent groups of girls and boys aged between 3 and 10 years and suffering from cerebral palsy was as shown

Table 6.8: The cognitive functioning of 17 elderly patients in a continuing care unit, measured before, and eight months after, the introduction of activity nursing, using a 13-question mental test score

PATIENT	SCORE		PATIENT	SCORE	
	Before	*After*		*Before*	*After*
Mr A	3	4	Mrs J	7	8
Mrs B	1	n/r	Mr K	6	3
Mr C	7	1	Mr L	10	12
Mrs D	6	9	Miss M	5	4
Mr E	12	13	Mrs N	3	4
Mrs F	12	12	Mrs O	4	5
Mrs G	7	8	Mrs P	9	11
Mrs H	7	7	Mrs Q	11	11
Mrs I	7	8			

Table 6.9: Daily energy intake of boys and girls aged between 3 and 10 years suffering from cerebral palsy

ENERGY INTAKE (MJ per day)	
Girls	*Boys*
6.3	6.7
7.3	6.4
10.0	7.3
11.5	5.6
8.4	10.2
5.2	8.0
8.7	

in Table 6.9. Use a suitable test to determine whether or not there is a significant difference in the median energy intakes of all such boys and girls. Explain why you have chosen the test you have used rather than a two-sample t test.

6.4 In an evaluation of a new device for the treatment of urinary incontinence, six women were treated with the device[11]. Before and after 10 treatments each patient completed a urinary frequency chart which contained information on bladder capacity. Frequency of urination (times per day) and bladder capacity (ml) values before and after treatment are shown in Table 6.10. Use a suitable test, explaining the reasons for your choice, to determine whether or not there was a significant improvement in each of the two measures.

6.5 Recalculate chi-squared and an exact p-value for the health promotion education data in Table 6.6.

Table 6.10: Frequency of urination (times per day) and bladder capacity (ml) of six women before and after treatment with a new device for dealing with urinary incontinence

PATIENT	FREQUENCY OF URINATION (times per day)		BLADDER CAPACITY (ml)	
	Before	*After*	*Before*	*After*
1	8.1	6.1	149	189
2	8.6	6.8	217	155
3	6.4	4.6	193	283
4	8.2	5.6	203	250
5	7.4	5.2	296	460
6	8.0	4.6	87	200

REFERENCES

1. BEBC (1995) *Evaluating a Nursing-Led Service: An Interim Report*. P.O. Box 1496, Parkstone, Poole, BH12 3YD.
2. Smith, D. *et al.* (1995) Adults with a history of child sexual abuse: evaluation of a pilot therapy service. *BMJ*, **310**, 1175–8.
3. Hawton K. *et al.* (1995) Suicide in Oxford University Students. *British Journal of Psychiatry*, **166**, 44–6.
4. Altman, D. G. (1991) *Practical Statistics for Medical Research*. London: Chapman & Hall (for a lucid interpretation of *U*).
5. Koh, P. and Thomas, V. J. (1994) Patient-controlled analgesia (PCA): does time saved by PCA improve patient satisfaction with nursing care? *Journal of Advanced Nursing*, **20**, 61–70.
6. Michinson, S. (1995) A review of the health promotion and health education beliefs of traditional and Project 2000 student nurses. *Journal of Advanced Nursing*, **21**, 356–63.
7. Altman, D. G. (1991) *Practical Statistics for Medical Research*. London: Chapman & Hall (for a description of chi-squared for trend).
8. Cotton, J. R. *et al.* (1994) Dietary fat and appetite: similarities and differences in the satiating effect of meals supplemented with either fat or carbohydrate. *Journal of Human Nutrition and Dietetics*, **7**, 11–24.
9. Turner, P. (1993) Activity nursing and the changes in the quality of life of elderly patients: a semi-quantitative study. *Journal of Advanced Nursing*, **18**, 1727–33.
10. Reilly, J. J. *et al.* (1994) Energy intakes of children with cerebral palsy. *Journal of Human Nutrition and Dietetics*, **7**, 95–103.
11. Cawley, D. M. and Hendriks, O. (1992) Evaluation of the Endomed CV405 as a treatment for urinary incontinence. *Physiotherapy*, **78**, 496–7.

7

HYPOTHESIS TESTS: METRIC VARIABLES

❑ THE ONE-SAMPLE t TEST ❑ THE t TEST FOR TWO MATCHED SAMPLES ❑ THE t TEST FOR TWO INDEPENDENT SAMPLES ❑ DOING THE t TESTS WITH A COMPUTER ❑

PARAMETRIC OR BUST

In the previous chapter we looked at three non-parametric tests we can use with data that are at least ordinal. The sign test and the Wilcoxon test are used for either a single sample, or for two *matched* groups; and the Mann–Whitney test for two *independent* groups. If the data are metric, you will be able to use the t test (which I will describe in a moment) provided that you are also sure that the distribution is *Normal*. In other words, the t test is a *parametric* procedure. In these circumstances, the t test is more powerful than either of the two non-parametric tests. If the data are metric but you are not sure about the shape of the distribution, then you should play safe and use the appropriate non-parametric test.

THE ONE-SAMPLE *t* TEST

- *Used with:* a single sample with metric data which must be Normally distributed.
- *Used to test:* that the population mean μ of the variable in question is equal to some hypothesised value k, i.e.

  H_0: $\mu = k$
  H_1: $\mu \neq k$.

Brief Description of the One-sample t Test

In Chapter 6 we discovered that the term:

$$\Sigma \frac{(O - E)^2}{E}$$

has a χ^2 (chi-squared) distribution, which we could use to determine the probability of getting any particular *observed* value. It so happens that the term:

$$\frac{\bar{x} - k}{\text{s.e.}(\bar{x})}$$

has a t distribution (which we have already met in Chapter 2) and we can use the above expression to determine the probability of getting any particular $\bar{x}$ and hence calculate the corresponding p-value. If you turn back to page 14 of Chapter 2 you will see that s.e.($\bar{x}$) is the standard error of the sample mean and equals $\frac{s}{\sqrt{n}}$. Once we know the p-value we can decide whether or not to reject H_0. Hopefully you won't have to do any of this by hand, but just in case, and to illustrate the general idea of the t test, I'll do one simple example.

Imagine that you walk down to casualty one day to have a word with your friend Vlad, who now works there as a reception administrator. You notice that a large sign on the wall says "Average waiting time is now 30 minutes". After a glance at Vlad's log book which records the time when a patient first arrives in casualty and the time when they are first seen by a nurse or doctor, you jot down the time spent waiting by the last 20 patients (figures shown in Table 7.1) and just for the hell of it you bet Vlad that the mean waiting time of *all* arriving patients is *not* 30 minutes. This seems like a bet you can't lose, but you have momentarily forgotten that just because the sample mean may not be 30 minutes, it doesn't follow that the population mean cannot equal 30 minutes.

Table 7.1: Time spent waiting by 20 consecutive arrivals in a casualty department (minutes)

20	30	45	40	15	50	30	25	30	50	80	20	10	50	60	25	20	20	35	60

Vlad's null and your alternative hypotheses are:

H_0: $\mu = 30$ mins
H_1: $\mu \neq 30$ mins.

To test the null hypothesis we need to calculate the value of $\frac{\bar{x} - k}{\text{s.e.}(\bar{x})}$, which means calculating the sample mean and the sample s.d. Using a calculator these are: $\bar{x} = 35.75$ mins and $s = 18.16$ mins. Therefore:

$$\frac{\bar{x} - k}{\text{s.e.}(\bar{x})} = \frac{\bar{x} - k}{\frac{s}{\sqrt{n}}} = \frac{35.75 - 30}{\frac{18.16}{\sqrt{20}}} = 1.416.$$

The next step is to find the p-value using the t distribution values in Table A2. The number of degrees of freedom (i.e. the row of Table A2 we will use) is $(n - 1) = 20 - 1 = 19$. In row 19 of Table A2 we look for the largest value *below* our observed value of 1.416. This value is 1.328 in the 0.20 column, so we can say that the p-value is less than 0.20 but more than 0.10. If we had chosen an $\alpha = 0.05$, we would not therefore be able to reject H_0. To be significant at $\alpha = 0.05$ with d.f. = 19 we would have needed a calculated value of more than 2.093. For significance at $\alpha = 0.01$, the value would have to be more than 2.861. So we have to accept H_0, that the mean waiting time probably is 30 minutes and Vlad wins the bet (again!).

Using a Computer to Do the One-sample *t* Test

Using Minitab for the One-sample t Test

If the data in the waiting time example above are entered into column c1, say, of the Minitab worksheet along with the following command sequence:

Stat
  Basic Statistics
    1-Sample t
      Select c1
        ⊙ **Test mean**
          30 (in Value box)
            OK

then the output in Figure 7.1 is produced. Minitab gives the same value of $t = 1.42$ as we calculated by hand above, and a more precise p-value of 0.17, confirming a non-rejection of H_0.

```
MTB > TTest 30.0 ''Waiting'';
SUBC> Alternative 0.

T-Test of the Mean

Test of mu = 30.00 vs mu not = 30.00

Variable      N     Mean    StDev   SE Mean      T    P-value
Waiting      20    35.75    18.16     4.06    1.42      0.17
```

Figure 7.1: Output from the Minitab one-sample t test for the A&E waiting time example, with H_0: $\mu = 30$

Using EPI for the One-sample t Test

EPI can be used to perform a one-sample t test. The test is accessed through the **Analysis** section in the menu, and then by typing either the **freq** or **means** commands with the appropriate variable name at the EPI prompt. EPI tests the null hypothesis that the population mean is zero.

SPSS cannot be used to perform a one-sample t test, but the descriptive command on the **Stat** menu will calculate s.e.($\bar{x}$), which helps a little.

THE *t* TEST WITH TWO MATCHED SAMPLES

- *Used with:* two matched samples, with metric data known to be Normally distributed.
- *Used to test:* that there is no significant difference between the population means of two matched populations. The null hypothesis is that:

$$H_0: \mu_1 - \mu_2 = 0$$

against the alternative hypothesis:

H_1: $\mu_1 - \mu_2 \neq 0$.

where μ_1 and μ_2 are the means of the two populations.

Brief Description of the Two-matched-sample t Test

I will illustrate this test using the data on six women treated with a new device to control urinary incontinence, which appeared in question 6.4 on the Wilcoxon rank sum test in the previous chapter. The reason for its inclusion there was that, with so small a sample, we could not be sure that bladder capacity in a population of such women would be Normally distributed and so we used a non-parametric test. Let us now assume that the distribution is known to be Normal and the parametric *t* test is therefore appropriate. For convenience the data on the bladder capacity (ml), before and after treatment with the device, is given again in Table 7.2.

Table 7.2: Bladder capacity (ml) of six women before and after treatment with a new device to control urinary incontinence

SUBJECT	BLADDER CAPACITY (ml)		AFTER – BEFORE
	Before	*After*	*d*
1	149	189	40
2	217	155	–62
3	193	283	90
4	203	250	47
5	296	460	164
6	87	200	113
Total			392

Mean difference $\bar{d} = 392/6 = 65.33$ ml

The null hypothesis is that there is no difference between population mean bladder capacity before the treatment, μ_{BEFORE}, and the mean capacity after, μ_{AFTER}. The alternative hypothesis is that there is a difference. The hypotheses are thus:

H_0: $\mu_{AFTER} - \mu_{BEFORE} = 0$
H_1: $\mu_{AFTER} - \mu_{BEFORE} \neq 0$.

The first step is to calculate the difference between the before and after scores. These difference values, labelled *d*, are shown in the last column of the table. The mean of these differences is calculated as $\bar{d} = 65.33$ ml, and the standard deviation $s_d = 77.21$ ml. Now it can be shown that the term:

$$\frac{\bar{d}}{s_d/\sqrt{n}}$$

where *n* is the number of matched pairs, has a *t* distribution, and so we can compare the value of this expression with the value of *t* from the *t* table and decide whether or not to reject H_0. If the value of the above expression is larger than the *t* table value we can reject H_0, otherwise we accept H_0. In this example:

$$\frac{\bar{d}}{s_d/\sqrt{n}} = \frac{65.33}{\frac{77.21}{\sqrt{6}}} = \frac{65.33}{31.52} = 2.073.$$

There are six values of d and so the number of d.f. = $n - 1 = 5$. The critical value of t in row 5 and column 0.05 of the t table is 2.571. Since $2.073 < 2.571$ we cannot reject H_0. There does not appear to be a significant difference in the mean before-and-after bladder capacities as a result of treatment with this new device. This result may at first sight be rather surprising in view of the large increases in all but one of the pairs. However, the small sample size and the consequent small number of d.f. means that the critical value of t is correspondingly large, a reflection of the unreliability inherent in small samples.

Using a Computer to Do the Matched-pairs *t* Test

Using SPSS for the Matched-pairs t Test

If we enter the six values for the before-and-after bladder capacities into columns c1 and c2 of the SPSS datasheet, and follow the command sequence:

Statistics
  Compare Means
    Paired-Samples T test
      Select c1
        Select c2
          OK

we get the output shown in Figure 7.2. SPSS first calculates the mean, standard deviation and s.e. of the mean for the before-and-after values separately (as well as the correlation coefficient between them and its significance, something we will discuss in the next chapter). It then calculates the mean, s.d. and s.e. of the differences, along with the value of $\bar{d}/(s_d/\sqrt{n})$, which it labels "t-value" of 2.07. It then gives a p-value of 0.093, which with an $\alpha = 0.05$ means that we cannot reject H_0, as we have already seen above. Finally, SPSS calculates a 95% confidence interval for the difference between the before-and-after values of (–15.720 to 146.387). Notice that this interval includes zero, confirming the non-significance of the test result.

Using Minitab for the Matched-pairs t Test

To use Minitab to do the two-matched pairs t test it is first necessary to enter the before-and-after bladder capacity values in columns c1 and c2 and then to calculate a column of difference values using the **Mathematical Expressions** in the **Calc** menu (or the **let** command described previously), which is named "diff". The one-sample t test is then applied to these diff values, using the commands:

Stat
  Basic Statistics
    1-Sample t
      Select diff
        ⊙ **Test mean**
        OK

```
SPSS for MS WINDOWS Release 6.0

t-tests for paired samples

              Number of          2-tail
Variable        pairs     Corr    Sig      Mean        SD      SE of Mean
AFTER                                    256.1667   109.733     44.798
                  6      0.715  0.110
BEFORE                                   190.8333    69.898     28.536

    Paired Differences
    Mean          SD        SE of Mean    t-value    df   2-tail Sig
  65.3333      77.211        31.521        2.07      5      0.093

95% C.I. (-15.720, 146.387)
```

Figure 7.2: Output from the SPSS two-matched-pairs t test for the before-and-after bladder capacities of six women with urinary incontinence

The output shown in Figure 7.3 is then produced. These results from Minitab are identical to that from SPSS, confirming that the difference in mean bladder capacity is not significant. Notice that Minitab also automatically produces a confidence interval for the difference in population means.

```
MTB > TwoSample 95.0 ''Before'' ''After'';
SUBC> Alternative 0.

Two Sample T-Test and Confidence Interval

Twosample T for Before vs After

           N    Mean     StDev    SE Mean
Before     6    190.8     69.9       29
After      6    256      110         45

95% C.I. for mu Before - mu After: (-188, 57)

T-Test mu Before = mu After (vs not =): T = -1.23 P = 0.25 DF = 8

MTB > Name c3 = ''diff''
MTB > Let ''diff'' = ''Before'' - ''After''
MTB > TTest 0.0 ''diff'';
SUBC> Alternative 0.

T-Test of the Mean

Test of mu = 0.0 vs mu not = 0.0

Variable   N    Mean     StDev    SE Mean       T       P-Value
diff       6    -65.3    77.2      31.5       -2.07      0.093
```

Figure 7.3: Output from the Minitab two-matched-pairs t test for the before-and-after bladder capacities of six women with urinary incontinence

Using Excel for the Matched-pairs t Test

We can use the bladder capacity data in Table 7.2 to illustrate the use of Excel. If the data are entered into the spreadsheet in columns A and B with the first rows labelled "before" and "after", and these two columns selected, then the following commands will produce the output in Figure 7.4:

Tools
  Data Analysis
    t-Test: Paired Two-Sample for Means
      OK
        Type A1:A7 in Variable 1 Input Range box
          Type B1:B7 in the Variable 2 Input Range box
            Type C1:D7 in the Output Range box
              ☒ Labels
                Accept α 0.05
                  Accept Hypothesised Mean Difference 0
                    OK

Labels ☒ ensures that the values in the first row will be used as labels. As can be seen, the output from Excel is quite comprehensive and reasonably self-explanatory. The same two-tailed p-value of 0.0929 as SPSS (0.093) is

To do the matched-pairs t test

	Before	After
Mean	190.8333333	256.1667
Variance	4885.766667	12041.37
Observations	6	6
Pearson Correlation	0.714826744	
Pooled Variance	5482.833333	
Hypothesized Mean Difference	0	
df	5	
t	−2.072687756	
P(T <= t) one-tail	0.046460772	
t Critical one-tail	2.015049176	
P(T <= t) two-tail	0.092921545	
t Critical two-tail	2.570577635	

Figure 7.4: Output from Excel for the matched-pairs *t* test on bladder capacity data (Table 7.2)

calculated, confirming acceptance of the null hypothesis of no difference between before-and-after bladder capacities.

THE *t* TEST WITH TWO INDEPENDENT SAMPLES

- *Used with:* two independent samples, with metric data known to be Normally distributed.
- *Used to test:* that there is no significant difference between the population means of two populations. The null hypothesis is that:

$$H_0: \mu_1 - \mu_2 = 0$$

against the alternative hypothesis:

$$H_1: \mu_1 - \mu_2 \neq 0.$$

where μ_1 and μ_2 are the means of the two populations. An additional requirement of the test is that the standard deviations of the two populations are equal (or approximately so). This last assumption is known as the assumption of *homoskedastic variances.*

Brief Description of the Two-matched-sample t Test

The algebra needed to describe this test is a bit complicated. You may want to omit this section first time round and go directly to the section below describing the use of a computer for the test. It can be shown that the following expression has a *t* distribution:

$$\frac{\bar{x}_1 - \bar{x}_2}{\sqrt{\dfrac{s_p{}^2}{n_1} + \dfrac{s_p{}^2}{n_2}}}$$

where $\bar{x}_1$ and $\bar{x}_2$ are the sample means from the first and second samples and n_1 and n_2 are the two sample sizes (the denominator of this expression is in fact the standard error of the sampling distribution of the difference in sample means). The term s_p^2 is known as the *pooled variance*, and is given by the expression:

$$s_p^2 = \frac{(n_1 - 1)s_1^2 + (n_2 - 1)s_2^2}{n_1 + n_2 - 2}$$

where s_1^2 and s_2^2 are the squares of the two sample standard deviations. To perform the test we first calculate the value of s_p^2 and then of the expression above which has the t distribution. We then compare this value with the critical value from the t table in Table A2. The number of d.f., i.e. the row we use in the t table, is equal to $(n_1 + n_2 - 2)$.

To illustrate the test, we can use the data in question 6.3 relating to the daily energy intake (in MJ) of two independent samples of n_1 = seven girls and n_2 = six boys suffering from cerebral palsy. Using a calculator gives $\bar{x}_{girls}$ = 8.2 MJ, $\bar{x}_{boys}$ = 7.367 MJ, s_{girls} = 2.154 MJ and s_{boys} = 1.608 MJ. The pooled variance is therefore:

$$s_p^2 = \frac{(7 - 1) \times 2.154^2 + (6 - 1) \times 1.608^2}{7 + 6 - 2} = 3.706$$

and

$$\frac{\bar{x}_1 - \bar{x}_2}{\sqrt{\dfrac{s_p^2}{n_1} + \dfrac{s_p^2}{n_2}}} = \frac{8.200 - 7.367}{\sqrt{\dfrac{3.706}{7} + \dfrac{3.706}{6}}} = \frac{0.833}{1.071} = 0.778.$$

In Table A2, with d.f. = 7 + 6 – 2 = 11, i.e. looking on row 11, the critical value of t is 2.201 with $\alpha = 0.05$. Since $0.778 < 2.201$ we cannot reject H_0, and it seems that there is no significant difference in the population mean daily energy intakes of such boys and girls. In practice, with samples as small as this, it would be better to use the Mann–Whitney test as we did in Chapter 8.

Using a Computer to Do the Two-independent-samples *t* Test

Using SPSS for the Two-independent-samples t Test

We can use the daily energy intake of children with cerebral palsy to demonstrate the use of SPSS, which assumes that the data are all in one column (c1), and are distinguished by a grouping variable in the adjacent column (c2; with 0 = boys and 1 = girls). The following command sequence will produce the output shown in Figure 7.5:

Statistics
  Compare Means
    Independent-Samples T Test
      Select Test Variable (c1)
        Select Grouping Variable (c2)
          Define Groups
            1 (Group 1)

```
SPSS for MS WINDOWS Release 6.0

t-tests for independent samples of SEX  sex

               Number
Variable     of cases      Mean        SD       SE of Mean
MJ
Girls            7        8.2000     2.154        0.814
Boys             6        7.3667     1.608        0.657

Mean Difference = 0.8333

Levene's Test for Equality of Variances F = 0.660 P = 0.434

t-test for Equality of Means          95%

Variances     t-value      df       2-Tail     SE of        CI for DIff
                                     Sig       Diff
Equal          0.78       11        0.453      1.071    (-1.525, 3.191)
Unequal        0.80       10.84     0.443      1.046    (-1.469, 3.136)
```

Figure 7.5: Output from the SPSS two-independent-sample t test for the mean daily energy intake of six boys and seven girls aged between 3 and 10 years, suffering from cerebral palsy

0 (Group 2)
Continue
OK

The first part of the SPSS output confirms our by-hand values for s_{girls}, s_{boys}, $\bar{x}_{girls}$ and $\bar{x}_{boys}$, and the value of 0.833 for the point estimate of the difference between the two means. The result of Levene's test of the equality of the two population variances is given next and they are found to be not significantly different (p-value = 0.434). The results of the t test for both equal and unequal population variances is then given. For equal variances (our assumption), the p-value = 0.453 supports our by-hand acceptance of H_0, as does the test when unequal variances are assumed (p-value = 0.443). Notice the s.e. of the difference of 1.071 is also the same as our by-hand value.

Using Excel for the Two-independent-samples t Test

If the cerebral palsy data are entered into columns A and B of the Excel spreadsheet, then the following commands will get Excel to do the independent pairs t test:

Tools
Analysis
t-Test: Two-Sample Assuming Equal Variance
OK
Type A1:A8 in Variable 1 Input Range box
Type B1:B7 in the Variable 2 Input Range box
Type C1:D10 in the Output Range box

☒ **Labels**
Accept α 0.05
Accept Hypothesised Mean Difference 0
OK

These commands assume that the distribution of energy intakes for boys and girls has an equal variance (spread). If this cannot be assumed then the unequal variance option should be chosen. The output from Excel is shown in Figure 7.6.

t-Test: Two-Sample Assuming Equal Variances

	Girls	Boys
Mean	8.2	7.366667
Variance	4.64	2.586667
Observations	7	6
Pooled Variance	3.706667	
Hypothesized Mean Difference	0	
df	11	
t	0.778001	
P(T<=t) one-tail	0.226489	
t Critical one-tail	1.795884	
P(T<=t) two-tail	0.452978	
t Critical two-tail	2.200986	

Figure 7.6: Output from the Excel two-independent-sample *t* test for the mean daily energy intake of six boys and seven girls aged between 3 and 10 years, suffering from cerebral palsy

The output is reasonably self-explanatory. The two-tailed *p*-value of 0.452978 is the same as that given by SPSS (0.453), and confirms that there is no significant difference in the mean daily energy intakes of such boys and girls.

Using Minitab for the Two-independent-samples t Test

The girls and boys data were entered in columns c1 and c2. Figure 7.7 is the output produced using the following commands:

Stat
Basic Statistics
2-Sample t
⊙ **Samples in different columns**
(Click in "First" box)
Select c1
Select c2
☒**Assume equal variances**
OK

The output produced by Minitab is very similar to the SPSS output, except that Minitab, because we chose the equal variances box, does not do a test of this assumption, and gives only equal variance output. Minitab calculates a *p*-value of 0.45, similar to that produced by Excel and SPSS.

```
MTB > TwoSample 95.0 ''Girls'' ''Boys'';
SUBC> Alternative 0;
SUBC> Pooled

Two Sample T-Test and Confidence Interval

Twosample T for Girls vs Boys

              N       Mean      StDev    SE Mean
Girls         7       8.20       2.15       0.81
Boys          6       7.37       1.61       0.66

95% C.I. for mu Girls - mu Boys: (-1.52, 3.19)

T-Test mu Girls = mu Boys (vs not =): T = 0.78 P = 0.45 DF = 11

Both use Pooled StDev = 1.93
```

Figure 7.7: Output from the Minitab two-independent-sample t test for the mean daily energy intake of six boys and seven girls aged between 3 and 10 years, suffering from cerebral palsy

We cannot go further with hypothesis tests in this book. Procedures for tests of metric variables when there are more than two samples can be dealt with using either a technique known as analysis of variance, or the equivalent (and more versatile) multiple regression approach.

SUMMARY

This chapter has described the t test for metric data, both for single samples and for two either matched or independent samples. The t test is the most powerful hypothesis test available in these circumstances, but is a parametric test and as such requires the assumption of Normally distributed population variables. If this assumption cannot be met, then an equivalent non-parametric test should be used. However, the use of confidence intervals is much to be preferred if this is at all possible.

In the last chapter of this book I want to have a brief look at what are known as measures of association or correlation.

EXERCISES

7.1 A sample of 25 patients included in a study of retinopathy and insulin-dependent diabetes, had their body mass index measured (data in Table 7.3). Test the null hypothesis that the population mean body mass index of such individuals is equal to 22.5. Use $\alpha = 0.01$.

7.2 Use the two-matched-samples t test to determine whether there is a significant difference in the mean length of fit between the first and fourth seizures in the patients having ECT (data in Table 2.7).

Table 7.3: Body mass index (kg/m^2) of a sample of 25 patients included in a study of retinopathy and insulin-dependent diabetes

23.9	20.7	27.3	25.6	21.3	27.6	24.3	27.1	21.5	20.9	22.6	21.9	26.0	22.9
25.5	26.8	27.6	23.8	22.1	22.5	20.1	24.5	25.1	18.7	23.5			

7.3 Use the data in Table 3.5, the waiting time for a hernia repair in two separate hospitals to perform a two-independent-samples t test that there is no difference in waiting times.

8

CORRELATION

❑ THE IDEA OF ASSOCIATION ❑ CORRELATION ❑ ASSOCIATION WITH NOMINAL DATA ❑ THE SCATTER DIAGRAM ❑ MEASURING ASSOCIATION WITH ORDINAL DATA – SPEARMAN'S RHO ❑ ASSOCIATION WITH METRIC VARIABLES – PEARSON'S *r* ❑ GETTING CORRELATION COEFFICIENTS WITH A COMPUTER ❑

WHAT IS THIS THING CALLED ASSOCIATION?

In the preceding chapters we have seen that one common application of statistical inference is to compare the values of various population parameters; for example, the mean age of male and female student suicides, the median before-and-after Oswestry Mobility Scale scores, the proportion of school children who were aware of illicit drugs in 1984 compared with 1994, and so on. In all of these examples we were working with a single variable; e.g. age, or Oswestry scores, or the proportion of drug-aware school children. However, we can also use inferential statistics to investigate the *relationships* between two or more variables. One area of particular interest is the examination of the possible *association* between two variables.

The word "association" in this context means the degree to which two variables have a tendency to "go together" or "move together". For example, high levels of cigarette smoking in populations of individuals tend to go together (to be associated) with high incidences of lung cancer among those same populations. Similarly, high levels of low-density-lipoprotein cholesterol tend to go with high levels of coronary heart disease. Alternatively, low levels of blood pressure are thought to be associated with *high* levels of depressive illness, and low socio-economic status with *higher* levels of general ill-health. As these examples show, there are two types of association:

- *Positive association:* high values of one variable tend to be associated with high values of the other variable, and vice versa (low values with low values). For example, smoking and lung cancer.
- *Negative association:* high values of one variable tend to be associated with low values of the other variable, and vice versa (low values with high values). For example, socio-economic status and ill-health.

The Scatter Diagram

A simple yet very effective way of examining the degree of association is with a scatter diagram or scattergram. In the scatter diagram, each pair of values of the

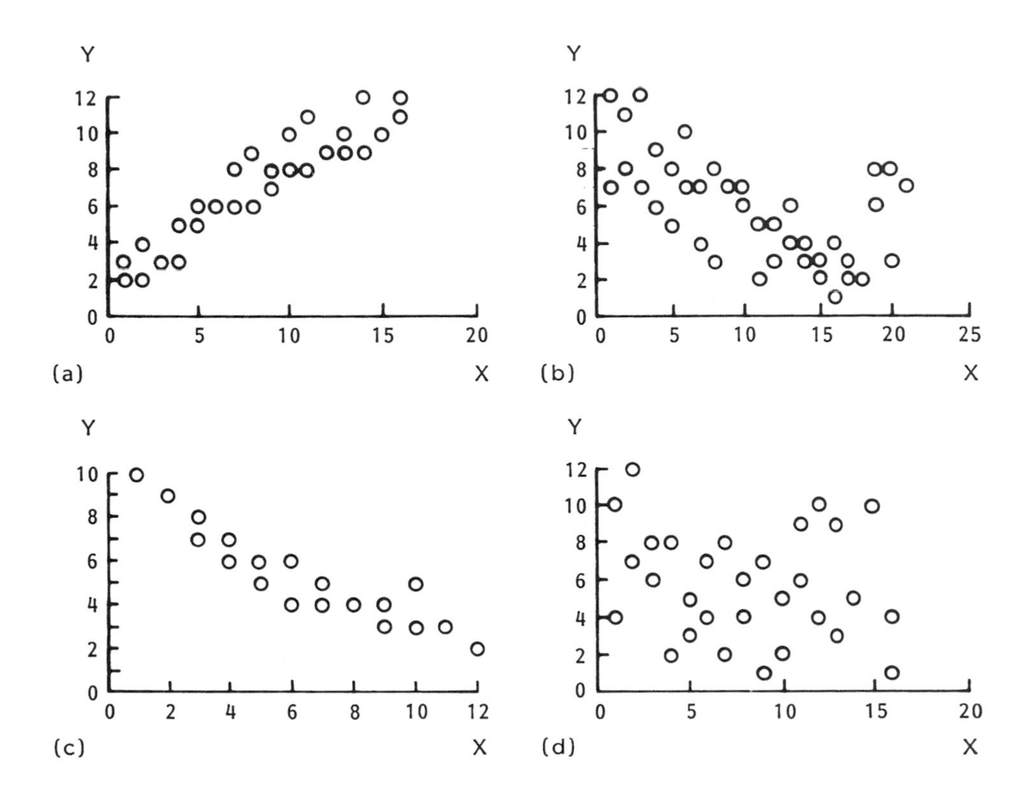

Figure 8.1: Varying strengths of association between two variables: (a) strong positive association; (b) moderate negative association; (c) strong negative association; (d) very weak association that might be positive or negative

two variables X and Y are plotted with the value of the first variable on the horizontal (or X) axis, and the value of the second on the vertical (or Y) axis. The "shape" of the scatter gives a general indication of the strength of the association. Figure 8.1 shows some examples illustrating various strengths of positive and negative association between two variables.

Scatter plots can easily be produced using a computer. In Minitab the required commands are **Graph**, **Plot**. Then select the columns of data (the first column of data is plotted as the vertical or Y axis, the second as the X axis) and click on **OK**. With SPSS, choose **Graph**, **Scatter**, **Simple**. Then select the required variables and click on **OK**. With Excel, enter the sample data into two columns of the spreadsheet, columns A and B say, select these columns, then click on the **Chart Wizard** and allow Chart Wizard to guide you through the necessary steps (Microsoft Word's **Chart** does the same thing).

The scatter diagram provides a quick way of assessing the strength and direction of any association between two variables. However, it is not very precise. It's rather like using a histogram to guess the mean of a set of sample values. More useful would be an exact numeric measure of the strength of the association. This is what the correlation coefficient provides.

The Correlation Coefficient

The sample correlation coefficient is a single numeric measure which is calculated using all of the sample values of the two variables (we'll see how shortly). It has a range of possible values from –1.0 through 0, to +1.0. These values are interpreted as follows.

- A value of +1.0 means that the two variables are *perfectly* positively associated. For example, an increase in cigarette smoking would be matched by an exactly equivalent increase in the incidence of lung cancer. If we were to plot a scatter diagram of the pairs of values we would get the scatter in Figure 8.2(a). In other words, the values lie exactly on a straight line, sloping up from left to right.
- A value of 0 means that the two variables are *not* associated at all. For example, if the value of one variable were to increase, the value of the second variable might increase, it might decrease, or it might remain unchanged. A scatter diagram would show a random "cloud" of values with no discernible shape, as in Figure 8.2(b).
- A value of –1.0 means that the two variables are *perfectly* negatively associated. For example, a decrease in blood pressure would be matched by an exactly equivalent increase in a depression scale score. If we were to plot a scatter diagram of the pairs of values we would get the diagram in Figure 8.2(c). In other words, the values lie exactly on a straight line, sloping down from left to right.

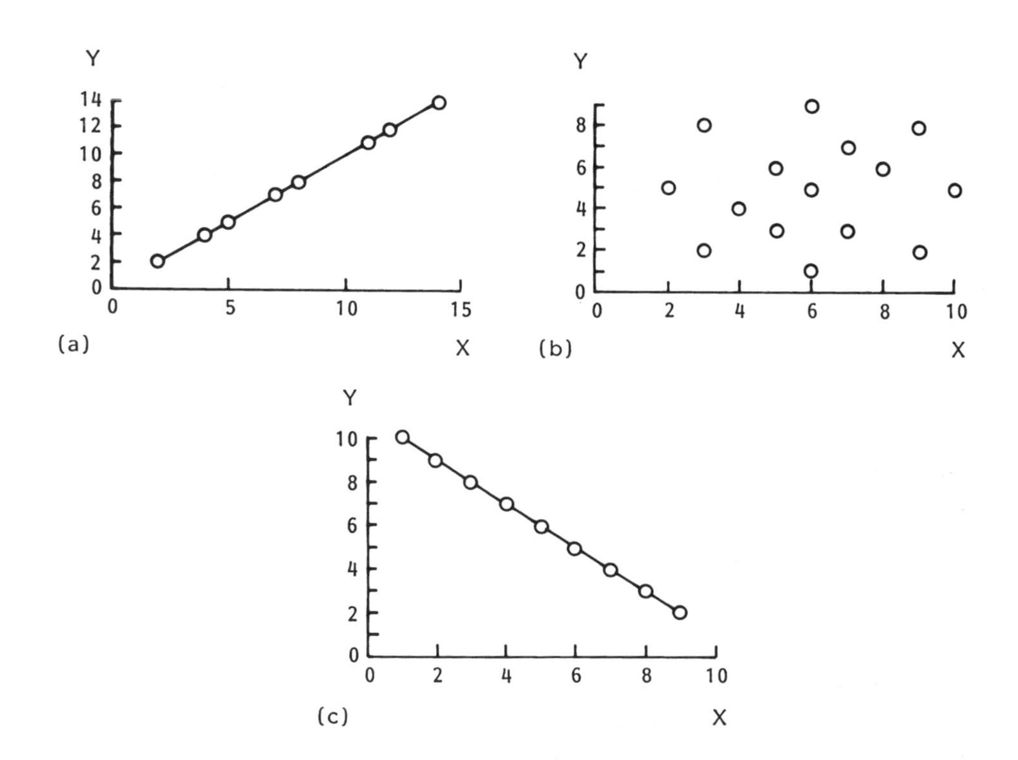

Figure 8.2: Scatter diagrams illustrating (a) perfect positive association, (b) zero correlation, and (c) perfect negative correlation

In practice, with real data, *perfect* correlation, either negative or positive, will never occur.

The actual magnitude and sign of a *sample* correlation coefficient is thus an *estimate* of the strength and direction of the association between the two variables in the *population*. In other words, the sample correlation coefficient is an estimate of the population correlation coefficient, ρ (Greek rho, pronounced as *row* a boat) The closer the value of the sample correlation coefficient is to +1, the stronger the positive association between the two variables in the population. The closer the value of the sample correlation coefficient is to –1, the stronger the negative association. The closer the value to zero, the weaker that association is. Like any other point estimator, therefore, the sample correlation coefficient has a degree of uncertainty attached to it, caused by random sampling error. We can assess the probable value of the population correlation coefficient by calculating the corresponding confidence interval, using the sample correlation coefficient as our starting point.

Two important features of correlation coefficients must be noted before we go any further. First, they only measure the strength of what are called *linear* relationships between pairs of variables. In other words, a correlation coefficient measures how close to lying exactly on a straight line is the scatter of points. So "linear" means "straight line" in this context. It's possible for there to be a strong relationship between two variables, such as that shown in Figure 8.3, but which is not linear. A correlation coefficient for this scatter might have a value close to zero, which might at first sight indicate a weak or no association between the variables in question, but in fact is simply because the association is not linear and the scatter does not therefore lie close to a straight line.

The second point is that correlation does not imply "cause". That is, just because two variables are strongly correlated does not mean that one "causes" the other. For example, if low blood pressure is correlated with depressive illness, there is no assumption that low blood pressure is the cause of depressive illness, nor is a depressive illness a cause of low blood pressure. In the same way, there is no assumption within correlation that smoking causes lung cancer, no more than that lung cancer causes people to smoke. Of course one variable *may* cause another, but this does *not* follow from the fact that they are correlated. If we want to study

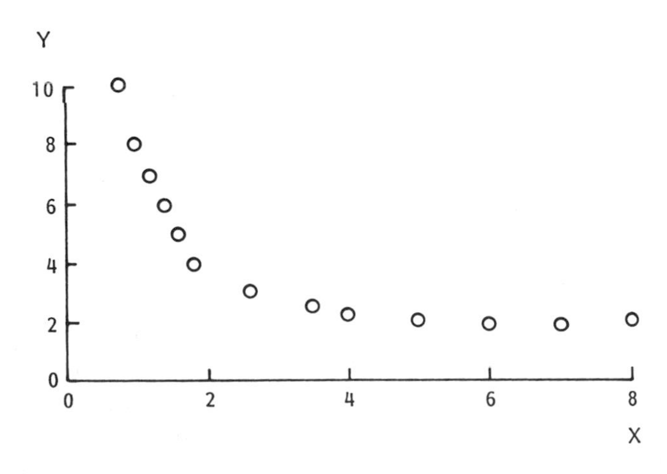

Figure 8.3: Scatter diagram for a strong but non-linear association between two variables

cause in the context of two (or more) variables we would have to turn to regression, for which there is no space in this book.

Having considered these preliminaries I want now to describe two widely used correlation coefficients. The first is used when both variables are metric, the second when either one or both variables are ordinal.

CORRELATION WITH TWO METRIC VARIABLES: PEARSON'S *r*

When *both* the variables are measured with metric data the strength of the association between them is determined using *Pearson's* correlation coefficient, denoted r. This is calculated by substituting the sample values of X and Y into the following rather formidable looking expression (although it looks worse than it is):

$$r = \frac{\Sigma XY - \dfrac{\Sigma X \Sigma Y}{n}}{\sqrt{\left[\Sigma X^2 - \left(\dfrac{\Sigma X}{n}\right)^2\right]\left[\Sigma Y^2 - \left(\dfrac{\Sigma Y}{n}\right)^2\right]}}.$$

I think you would want to avoid working this out by hand if at all possible! Some calculators will do it, and SPSS, Minitab and Excel will calculate r. However, just in case, Table 8.1 is an example of how r is calculated using just three sample pairs of observations. The result of 0.204 shows a small positive association between the sample values of X and Y. However, without either a confidence interval or a hypothesis test we cannot say whether this sample value of r is indicative of a population correlation coefficient ρ significantly different from zero (and hence indicative of a significant association between the population variables). In general, the value of r will not equal zero, even when ρ is zero, because of sampling error.

Table 8.1: Calculating Pearson's r by hand

X	Y	$X \times Y$	X^2	Y^2
1	3	3	1	9
2	4	8	4	16
5	6	30	25	36
$\Sigma X = 8$	$\Sigma Y = 13$	$\Sigma XY = 41$	$\Sigma X^2 = 30$	$\Sigma Y^2 = 61$

$$r = \frac{41 - \dfrac{8 \times 13}{3}}{\sqrt{\left[30 - \left(\dfrac{8}{3}\right)^2\right]\left[61 - \left(\dfrac{13}{3}\right)^2\right]}} = \frac{6.333}{\sqrt{22.889 \times 42.222}} = \frac{6.333}{31.087} = 0.204$$

With a confidence interval we would want to know whether the interval included 0. If it did, this would imply a population correlation coefficient ρ between X and Y of zero. For example, an interval of (–0.078, 0.106) would imply a ρ not significantly different from zero. On the other hand, an interval that did not include zero, say, for example, an interval of (0.114, 0.205), would imply that ρ was significantly different from zero. Unfortunately, the calculations required for confidence intervals for r are quite awkward and will be omitted. Interested readers will find these treated elsewhere[1].

With a hypothesis test, the procedure is easier. The null hypothesis would be: H_0: $\rho = 0$, the alternative hypothesis H_1: $\rho \neq 0$. The sample value of r has to be compared with a table of critical r values at the appropriate level of significance, α, and the decision made to accept or reject H_0. I will describe the hypothesis test procedure in detail in the example from practice below. However, as a rule of thumb, if $r > 2/\sqrt{n}$ then ρ is significantly different from zero, but this approximation only works for $n > 20$.

An Example from Practice

Researchers[2] investigating the low rate of coronary heart disease (CHD) in France compared with other developed countries with comparable dietary intake ("the French paradox"), examined alcohol, diet and mortality data in a sample of 21 developed countries. Table 8.2 contains data on the annual *per capita* consumption of wine ethanol (litres) and the age-adjusted mortality rate from CHD (deaths per 100 000 of the population) in 1988 in the countries in the study. Figure 8.4 is a scatter diagram of the values which shows a fairly strong negative association between the two variables, i.e. high levels of wine ethanol consumption seem to be associated with low CHD mortality rates, and vice versa.

Table 8.2: Annual *per capita* consumption of wine ethanol (litres) and the mortality rate from coronary heart disease (CHD) (deaths per 100 000 of the population) in 21 developed countries

COUNTRY	WINE ETHANOL (annual litres *per capita*)	CHD MORTALITY (rate per 100 000)
Australia	2.5	215
Austria	3.9	160
Belgium	2.9	120
Canada	2.4	190
Denmark	2.9	210
Finland	0.8	295
France	9.1	70
Iceland	0.8	200
Ireland	0.7	310
Israel	0.6	180
Italy	7.9	105
Japan	1.5	30
Netherlands	1.8	160
New Zealand	1.9	260
Norway	0.8	215
Spain	6.5	80
Sweden	1.6	200
Switzerland	5.8	110
UK	1.3	280
USA	1.2	170
W. Germany	2.7	165

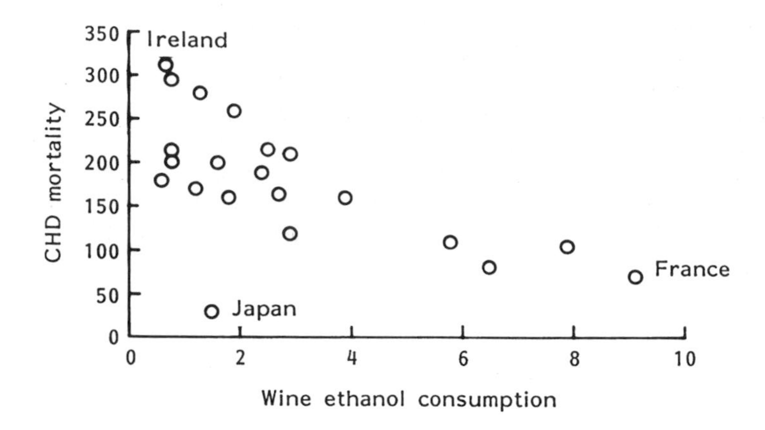

Figure 8.4: Scatter diagram showing the association between consumption of wine ethanol and CHD mortality rates (data in Table 8.1)

The authors calculated Pearson's r for these data to be –0.66, although they did not report whether this value is significant. As a matter of interest, they concluded that "The total mortality data demonstrate that the benefits of alcohol preventing CHD at a population level are essentially cancelled by increases in other causes of death and suggests there is a protective effect on total mortality only for light to moderate drinking".

Testing the Significance of *r*

Although the significance of r can easily be tested using a computer (see below), I will describe the by-hand procedure in case you ever need it, using the wine ethanol and CHD example above, for which $r = -0.66$ with $n = 21$. The hypotheses are:

H_0: $\rho = 0$
H_1: $\rho \neq 0$

The decision rule is that:

- if the sample value of r (ignoring the sign of r) is greater than the tabulated critical value, reject H_0;
- otherwise H_0 cannot be rejected.

The critical values of r are given in Table A8. The appropriate number of degrees of freedom (d.f.), i.e. the row we have to use, is $n - 2$, so in this example d.f. = 21 – 2 = 19. In the 0.05 column on row 19, the critical value is 0.4334. Since 0.66 > 0.433 we can reject H_0 and conclude that wine ethanol consumption and the CHD mortality rate are significantly negatively associated in the population.

Using a Computer to Get Pearson's Correlation Coefficient

I will use the wine ethanol/CHD example to illustrate the use of computers to calculate r, calculate confidence intervals and perform significance tests.

Using SPSS for Pearson's Coefficient

The data on wine ethanol consumption and CHD mortality rate from Table 8.2 are entered into columns 1 and 2, say, of the SPSS datasheet and named "wine" and "chd". The following commands will produce the output in Figure 8.5.

```
SPSS for MS WINDOWS Release 6.0

Correlation Coefficients

                 CHD          WINE
CHD            1.0000       -0.6348
               (21)          (21)
               P = .        P = 0.002
WINE          -0.6348        1.0000
               (21)          (21)
              P = 0.002      P = .

(Coefficient / (Cases) / 2-tailed Significance)

''.'' is printed if a coefficient cannot be computed
```

Figure 8.5: Output from SPSS for Pearson's correlation coefficient between wine consumption and CHD mortality rate (data in Table 8.2)

Statistics
 Correlation
  Bivariate
   Select chd
    Select wine
     ☒ Pearson
      OK

The value of r calculated by SPSS is –0.6348. The difference between this value and the value of 0.66 quoted by the authors in the example above is because I had to estimate the CHD values from a bar chart, the authors did not report the actual figures. SPSS calculates a p-value of 0.002 which, since this is less than 0.05, enables us to reject the null hypothesis of no significant association, as we were able to above.

Using Minitab for Pearson's Coefficient

To use Minitab to calculate r for the wine and CHD data entered in columns c1 and c2, the commands needed are:

Stat
  Basic Statistics
    Correlation
      Select c1
        Select c2
          OK

The output from Minitab is shown in Figure 8.6. The value for r is the same as with SPSS, but Minitab provides no further information, e.g. no p-value, which means that it is not possible to say whether this value of r is significant, without resorting to a table of critical values.

```
MTB > Correlation C2 C3.

Correlations (Pearson)

Correlation of C2 and C3 = –0.635
```

Figure 8.6: Output from Minitab for Pearson's correlation coefficient between wine consumption and CHD mortality rate (data in Table 8.2)

Using Excel for Pearson's Coefficient

The data on wine ethanol consumption and CHD is entered into columns A and B of the Excel spreadsheet. The first row of the columns *must* contain data labels, in this case "wine" and "chd". The columns are selected by clicking in cell A to select the first column and clicking on B, with the CTRL key pressed, to select the second column. The following commands will produce the output shown in Figure 8.7:

Tools
  Data Analysis
    Correlation
      Input Range: A1:B21
        Output Range: C1
          Grouped by: ⊙ Columns
            ☒ Labels in First Row
              OK

```
                    Wine     CHD
Wine                   1
CHD             -0.63577       1
```

Figure 8.7: Output from Excel for Pearson's correlation coefficient between wine ethanol consumption and CHD mortality rate (data in Table 8.2)

The output from Excel is spartan and shows only the value of Pearson's r of –0.63577 compared with SPSS's –0.6348 and Minitab's –0.635. No p-value is given, which means resorting to the table of significant r values to determine whether or not this value is significant.

Using CIA for Pearson's Coefficient

The CIA will calculate r and a confidence interval for r. The commands necessary are:

CHAPTER 5 – Regression and Correlation
3 : Correlation
1. Pearson's correlation coefficient

The program then asks for the sample size and for the data to be input. The output is shown in Figure 8.8. CIA produces a value for r of –0.635 and a 95% confidence interval of (–0.837 to –0.280). So we can be 95% certain that the population correlation coefficient lies somewhere between these two limits. The fact that this interval does not include zero confirms of course the non-zero significance of ρ.

```
SAMPLE SIZE : 21
PEARSON'S PRODUCT MOMENT CORRELATION COEFFICIENT : -.635

% CONFIDENCE REQUIRED : 95

-----------------------------------------

Standard Error of z-transformed Correlation Coefficient = 0.236
NORMAL Value = 1.96

95% CONFIDENCE INTERVAL FOR THE CORRELATION COEFFICIENT IS:
-0.837 TO -0.280

-----------------------------------------

Another level of confidence (Y/N)?
```

Figure 8.8: Output of CIA for Pearson's r and 95% confidence interval between wine ethanol consumption and CHD (data in Table 8.2)

CORRELATION WHEN ONE OR MORE VARIABLES ARE ORDINAL: SPEARMAN'S r_s

Pearson's r should be used only when *both* the variables are metric. If either one or both variables are ordinal then Spearman's rank correlation coefficient r_s is appropriate. Calculation of r_s is less awkward than r, and in case it has to be calculated by hand, the necessary steps are as follows:

- *Step 1:* Rank the scores of X into ascending order; label these R_X.
- *Step 2:* Do the same for the scores on Y; call these R_Y.
- *Step 3:* Calculate the difference between each pair of R_X and R_Y values; call these differences d.
- *Step 4:* Square each d value, to get a set of d^2 values.
- *Step 5:* Add these d^2 values to get Σd^2. Multiply by 6.
- *Step 6:* Square n, the sample size, to get n^2 and subtract from 1. Multiply the result by n.
- *Step 7:* Divide the value from Step 3 by the value from Step 4.
- *Step 8:* Subtract the value from Step 5 from 1. This is r_s.

Those more at home with algebraic expressions can substitute directly into the expression:

$$r_s = 1 - \frac{6 \times \Sigma d^2}{n(n^2 - 1)}$$

where n is the number of pairs of sample data.

To illustrate the use of Spearman's r_s we can use the data in Table 8.3, the first three columns of which record the subject number, the body mass index, and the score on the Eating Inventory Test (EIT), a measure of dietary restraint, of 16 subjects in an investigation into the relationship between dietary fat and appetite[3]. The scatter diagram for the sample data is displayed in Figure 8.9. This shows a generally undefined shape, with little sign of linearity, indicating weak (if any) association between the variables.

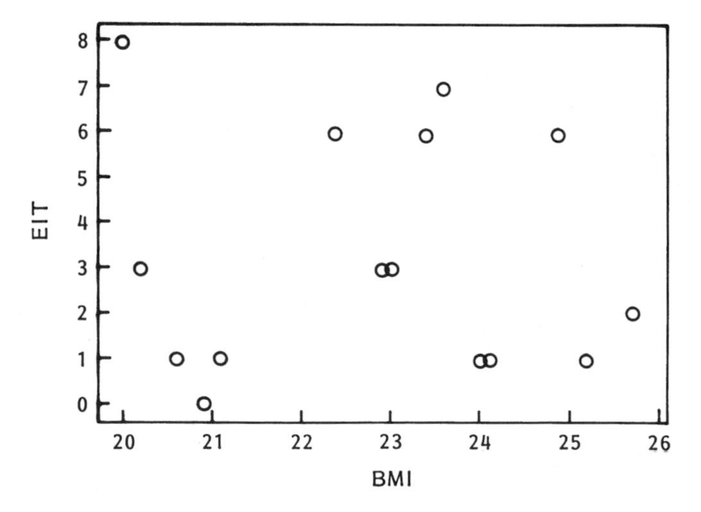

Figure 8.9: Scatter diagram for the BMI–EIT scores (data in Table 8.3)

Table 8.3: The body mass index BMI (kg/m²), and the score on the Eating Inventory Test (EIT), of 16 subjects in an investigation into the relationship between dietary fat and appetite

SUBJECT	BMI	EIT	R_{BMI}	R_{EIT}	d	d^2
1	20.6	1	3	4.5	−1.5	2.25
2	24.1	1	13	4.5	8.5	72.25
3	22.4	6	6	13	−7.0	49.00
4	20.2	3	2	10	−8.0	64.00
5	25.2	1	15	4.5	10.5	110.25
6	21.1	1	5	4.5	0.5	0.25
7	24.0	1	11.5	4.5	7.0	49.00
8	20.9	0	4	1	3.0	9.00
9	24.9	6	14	13	1.0	1.00
10	23.0	3	8	10	−2.0	4.00
11	22.9	3	7	10	−3.0	9.00
12	25.7	2	16	8	8.0	64.00
13	23.4	6	9	13	−4.0	16.00
14	24.0	1	11.5	4.5	7.0	49.00
15	23.6	7	10	15	−5.0	25.00
16	20.0	8	1	16	−15.0	225.00

The ranked BMI and EIT scores are shown in columns R_{BMI} and R_{EIT}, of Table 8.3 (using the Minitab **Rank** command from the **Manip** menu) followed by columns of d and d^2 values. The sum of the d^2 values is $\Sigma d^2 = 749.0$. With $n = 16$, $n^2 = 256$, and Spearman's r_s is therefore:

$$t_s = 1 - \frac{6 \times 749.0}{16(256-1)} = 1 - \frac{4494}{4080} = 1 - 1.101 = -0.101.$$

This result indicates a small negative association between BMI and EIT scores in this sample, and confirms the observation made from the scatter diagram of little or no association between the variables. To determine conclusively whether this value of r_s is indicative of a significant correlation in the population between BMI and EIT scores, we will need to refer to Table A9.

Table A9 contains critical values of r_s which must be exceeded for the result to be significant. In this case with $n = 16$, the critical value of r_s is 0.506 with $\alpha = 0.05$. Since 0.101 (we ignore the negative sign) does *not* exceed 0.506, we cannot reject the null hypothesis of no association between these two variables in the population.

Using a Computer to Get Spearman's Correlation Coefficient

We can use the above BMI/EIT example to illustrate the use of a computer to get r_s. Minitab does not calculate r_s directly, and the data have first to be ranked before Pearson's r is calculated on the ranked values (Pearson's r on ranked values is exactly the same as Spearman's r_s), which is all a bit too inconvenient if SPSS or CIA are available.

Using SPSS for Spearman's Coefficient

If the BMI and EIT data are entered into columns 1 and 2, say, and named, then the following commands will produce the output shown in Figure 8.10:

Statistics
  Correlation
    Bivariate
      Select BMI
        Select EIT
          ☒ **Spearman**
            ☐ **Pearson**
              OK

```
SPSS for MS WINDOWS Release 6.0

Spearman Correlation Coefficients

EIT     -0.1384
        N (16)
        Sig 0.609

        BMI

(Coefficient / (Cases) / 2-tailed Significance)

''.'' is printed if a coefficient cannot be computed
```

Figure 8.10: Output from SPSS for Spearman's r_s for the BMI–EIT data (data in Table 8.3)

SPSS calculates a value of r_s = –0.1384 with a p-value (labelled SIG) of 0.609, confirming our suspicions from the evidence of the scatter diagram in Figure 8.8 that ρ_s is not significantly different from zero. In other words, BMI and EIT are not associated in this population.

Using CIA for Spearman's Coefficient

Using CIA to get a value for r_s and the corresponding confidence intervals requires more or less the same commands as for Pearson's r described above, except that Spearman's correlation coefficient is chosen instead of Pearson's. The necessary commands are:

CHAPTER 5 – Regression and Correlation
  3 : Correlation
    2. Spearman's correlation coefficient

The output from CIA for the BMI/EIT example is shown in Figure 8.11. CIA calculates a value for r_s of 0.0568, with a 95% confidence interval of (–0.452 to 537). I believe the different value for r_s between the by-hand value, SPSS and CIA is caused by the presence of many ties in the data and the different treatment accorded to them by the programs. Nonetheless, the CIA confidence interval, which includes zero, confirms the SPSS p-value result of no significant association between the variables.

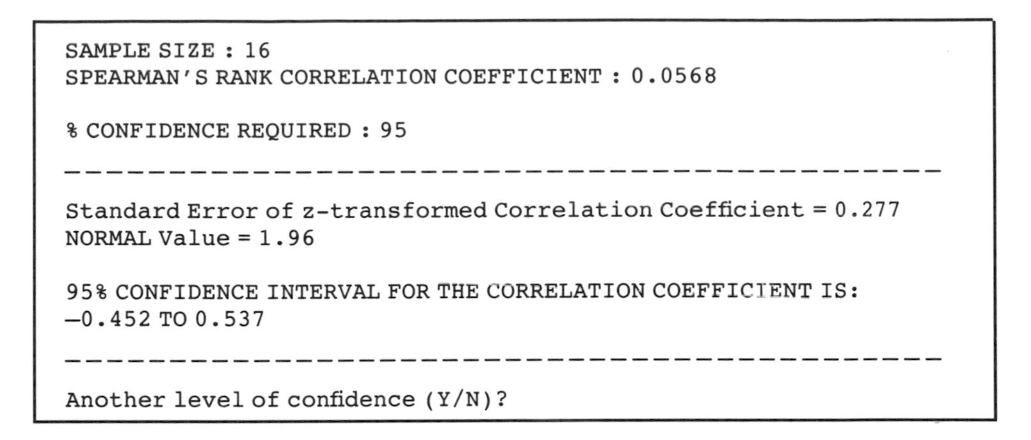

```
SAMPLE SIZE : 16
SPEARMAN'S RANK CORRELATION COEFFICIENT : 0.0568

% CONFIDENCE REQUIRED : 95
------------------------------------------
Standard Error of z-transformed Correlation Coefficient = 0.277
NORMAL Value = 1.96

95% CONFIDENCE INTERVAL FOR THE CORRELATION COEFFICIENT IS:
−0.452 TO 0.537
------------------------------------------
Another level of confidence (Y/N)?
```

Figure 8.11: Output of CIA for Spearman's r_s and 95% confidence interval for BMI and EIT scores (data in Table 8.3)

It is worth mentioning that Spearman's correlation coefficient is not well suited to data containing a substantial number of tied values. In these circumstances it might be better to use a measure of association known as Goodman and Kruskal's gamma statistic. Gamma can be calculated using the SPSS Crosstabulation procedure, but unfortunately space limitations do not permit further discussion of this measure[4]. Measures of association for nominal variables (not discussed here), e.g. Cramer's V, are also available in Crosstabulation.

Using Excel for Spearman's Coefficient

It would appear that Excel will not calculate r_s directly. However, the data could first be ranked and then Pearson's r calculated.

EXERCISES

8.1 Give some examples from your own experience of pairs of variables which are: (a) positively associated; (b) negatively associated.

8.2 How does (a) a correlation coefficient, and (b) a scatter diagram, indicate: (i) strong positive association; (ii) strong negative association; (iii) zero or weak association.

8.3 The data[5] in Table 8.4 refer to percentage mortality from aortic aneurysm and the number of patient admission episodes per year in a sample of 22 hospitals in the south of England 1992–93.

(a) Plot a scatter diagram of the data with mortality on the vertical axis. Comment on what is revealed about a possible association between the two variables.

(b) Calculate an appropriate measure of association and comment on the meaning of the result. Can it be said as a consequence of the variables being correlated that the number of episodes experienced by a hospital in some way affects the percentage mortality. Explain your answer.

Table 8.4: Percentage mortality for aortic aneurysm and the number of patient admission episodes per year in a sample of 22 hospitals in the south of England 1992–93

MORTALITY (%)	EPISODES PER YEAR
66	3
49	6
32	6
61	8
35	12
49	13
0	15
41	22
22	22
18	22
20	23
27	25
12	28
35	33
35	36
30	38
24	39
10	61

8.4 In the dietary study referred to earlier[3], a second experiment involving 12 subjects was carried out. Prior to their inclusion in the study the subjects had their BMI, EAT (Eating Attitude Test) score and EIT score measured. The results are shown in Table 8.5. Plot scatter diagrams and calculate appropriate measures of association for each possible pair of variables and comment on what is revealed.

Table 8.5: Measurements of BMI, EAT and EIT on 12 subjects in a dietary study

SUBJECT	BMI	EAT	EIT
1	25.6	3	5
2	20.0	7	8
3	22.5	3	6
4	23.6	3	6
5	24.6	4	8
6	20.3	0	3
7	21.8	2	1
8	20.3	4	6
9	24.5	0	1
10	24.8	3	7
11	24.9	3	2
12	23.5	0	2

REFERENCES

1. For example: Iman, R. L. and Conover, W. J. (1989) *Modern Business Statistics*. Chichester: John Wiley.
2. Criqui, M. H. and Ringel, B. L. (1994) Does diet or alcohol explain the French paradox? *The Lancet*, **344**, 1719–23.
3. Cotton, J. *et al.* (1994) Dietary fat and appetite: similarities and differences in the satiating effect of meals supplemented with either fat or carbohydrate. *Journal of Human Nutrition and Dietetics*, **7**, 11–24.
4. Siegel, S. and Castellan, N. J. (1988) *Nonparametric Statistics*. Maidenhead: McGraw-Hill (for an excellent and detailed survey of non-parametric measures of association, including Goodman and Kruskal's gamma).
5. McKee, M. and Hunter, D. (1995) Mortality league tables: do they inform or mislead? *Quality in Health Care*, **4**, 5–12.

APPENDIX

STATISTICAL TABLES

Table A1: Areas under the "standard Normal distribution" between $z = 0$ and chosen value of z. The area in the tail is equal to 0.5000 minus the value in the table

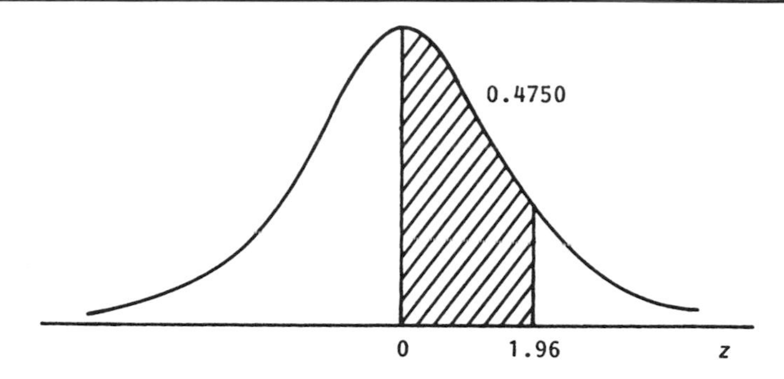

z	.00	.01	.02	.03	.04	.05	.06	.07	.08	.09
0.0	0.0000	0.0040	0.0080	0.0120	0.0160	0.0199	0.0239	0.0279	0.0319	0.0359
0.1	0.0398	0.0438	0.0478	0.0517	0.0557	0.0596	0.0636	0.0675	0.0714	0.0753
0.2	0.0793	0.0832	0.0871	0.0910	0.0948	0.0987	0.1026	0.1064	0.1103	0.1141
0.3	0.1179	0.1217	0.1255	0.1293	0.1331	0.1368	0.1406	0.1443	0.1480	0.1517
0.4	0.1554	0.1591	0.1628	0.1664	0.1700	0.1736	0.1772	0.1808	0.1844	0.1879
0.5	0.1915	0.1950	0.1985	0.2019	0.2054	0.2088	0.2123	0.2157	0.2190	0.2224
0.6	0.2257	0.2291	0.2324	0.2357	0.2389	0.2422	0.2454	0.2486	0.2517	0.2549
0.7	0.2580	0.2611	0.2642	0.2673	0.2704	0.2734	0.2764	0.2794	0.2823	0.2852
0.8	0.2881	0.2910	0.2939	0.2967	0.2995	0.3023	0.3051	0.3078	0.3106	0.3133
0.9	0.3159	0.3186	0.3212	0.3238	0.3264	0.3289	0.3315	0.3340	0.3365	0.3389
1.0	0.3413	0.3438	0.3461	0.3485	0.3508	0.3531	0.3554	0.3577	0.3599	0.3621
1.1	0.3643	0.3665	0.3686	0.3708	0.3729	0.3749	0.3770	0.3790	0.3810	0.3830
1.2	0.3849	0.3869	0.3888	0.3907	0.3925	0.3944	0.3962	0.3980	0.3997	0.4015
1.3	0.4032	0.4049	0.4066	0.4082	0.4099	0.4115	0.4131	0.4147	0.4162	0.4177
1.4	0.4192	0.4207	0.4222	0.4236	0.4251	0.4265	0.4279	0.4292	0.4306	0.4319
1.5	0.4332	0.4345	0.4357	0.4370	0.4382	0.4394	0.4406	0.4418	0.4429	0.4441
1.6	0.4452	0.4463	0.4474	0.4484	0.4495	0.4505	0.4515	0.4525	0.4535	0.4545
1.7	0.4554	0.4564	0.4573	0.4582	0.4591	0.4599	0.4608	0.4616	0.4625	0.4633
1.8	0.4641	0.4649	0.4656	0.4664	0.4671	0.4678	0.4686	0.4693	0.4699	0.4706
1.9	0.4713	0.4719	0.4726	0.4732	0.4738	0.4744	**0.4750**	0.4756	0.4761	0.4767
2.0	0.4772	0.4778	0.4783	0.4788	0.4793	0.4798	0.4803	0.4808	0.4812	0.4817
2.1	0.4821	0.4826	0.4830	0.4834	0.4838	0.4842	0.4846	0.4850	0.4854	0.4857
2.2	0.4861	0.4864	0.4868	0.4871	0.4875	0.4878	0.4881	0.4884	0.4887	0.4890
2.3	0.4893	0.4896	0.4898	0.4901	0.4904	0.4906	0.4909	0.4911	0.4913	0.4916
2.4	0.4918	0.4920	0.4922	0.4925	0.4927	0.4929	0.4931	0.4932	0.4934	0.4936
2.5	0.4938	0.4940	0.4941	0.4943	0.4945	0.4946	0.4948	0.4949	0.4951	0.4952
2.6	0.4953	0.4955	0.4956	0.4957	0.4959	0.4960	0.4961	0.4962	0.4963	0.4964
2.7	0.4965	0.4966	0.4967	0.4968	0.4969	0.4970	0.4971	0.4972	0.4973	0.4974
2.8	0.4974	0.4975	0.4976	0.4977	0.4977	0.4978	0.4979	0.4979	0.4980	0.4981
2.9	0.4981	0.4982	0.4982	0.4983	0.4984	0.4984	0.4985	0.4985	0.4986	0.4986
3.0	0.4987	0.4987	0.4987	0.4988	0.4988	0.4989	0.4989	0.4989	0.4990	0.4990

Table A2: The *t* distribution. For significance the sample *t* value must exceed the tabulated value below

Degrees of freedom	(80%) 0.2	(90%) 0.1	(95%) 0.05	(98%) 0.02	(99%) 0.01	(99.9%) 0.001
1	3.078	6.314	12.706	31.821	63.657	636.619
2	1.886	2.920	4.303	6.965	9.925	31.599
3	1.638	2.353	3.182	4.541	5.841	12.924
4	1.533	2.132	2.776	3.747	4.604	8.610
5	1.476	2.015	2.571	3.365	4.032	6.869
6	1.440	1.943	2.447	3.143	3.707	5.959
7	1.415	1.895	2.365	2.998	3.499	5.408
8	1.397	1.860	2.306	2.896	3.355	5.041
9	1.383	1.833	2.262	2.821	3.250	4.781
10	1.372	1.812	2.228	2.764	3.169	4.587
11	1.363	1.796	2.201	2.718	3.106	4.437
12	1.356	1.782	2.179	2.681	3.055	4.318
13	1.350	1.771	2.160	2.650	3.012	4.221
14	1.345	1.761	2.145	2.624	2.977	4.140
15	1.341	1.753	2.131	2.602	2.947	4.073
16	1.337	1.746	2.120	2.583	2.921	4.015
17	1.333	1.740	2.110	2.567	2.898	3.965
18	1.330	1.734	2.101	2.552	2.878	3.922
19	1.328	1.729	2.093	2.539	2.861	3.883
20	1.325	1.725	2.086	2.528	2.845	3.850
21	1.323	1.721	2.080	2.518	2.831	3.819
22	1.321	1.717	2.074	2.508	2.819	3.792
23	1.319	1.714	2.069	2.500	2.807	3.768
24	1.318	1.711	2.064	2.492	2.797	3.745
25	1.316	1.708	2.060	2.485	2.787	3.725
26	1.315	1.706	2.056	2.479	2.779	3.707
27	1.314	1.703	2.052	2.473	2.771	3.690
28	1.313	1.701	2.048	2.467	2.763	3.674
29	1.311	1.699	2.045	2.462	2.756	3.659
30	1.310	1.697	2.042	2.457	2.750	3.646
31	1.309	1.696	2.040	2.453	2.744	3.633
32	1.309	1.694	2.037	2.449	2.738	3.622
33	1.308	1.692	2.035	2.445	2.733	3.611
34	1.307	1.691	2.032	2.441	2.728	3.601
35	1.306	1.690	2.030	2.438	2.724	3.591
36	1.306	1.688	2.028	2.434	2.719	3.582
37	1.305	1.687	2.026	2.431	2.715	3.574
38	1.304	1.686	2.024	2.429	2.712	3.566
39	1.304	1.685	2.023	2.426	2.708	3.558
40	1.303	1.684	2.021	2.423	2.704	3.551
41	1.303	1.683	2.020	2.421	2.701	3.544
42	1.302	1.682	2.018	2.418	2.698	3.538
43	1.302	1.681	2.017	2.416	2.695	3.532
44	1.301	1.680	2.015	2.414	2.692	3.526
45	1.301	1.679	2.014	2.412	2.690	3.520

Table A2: (*continued*)

Degrees of freedom	(80%) 0.2	(90%) 0.1	(95%) 0.05	(98%) 0.02	(99%) (0.01	(99.9%) 0.001
46	1.300	1.679	2.013	2.410	2.687	3.515
47	1.300	1.678	2.012	2.408	2.685	3.510
48	1.299	1.677	2.011	2.407	2.682	3.505
49	1.299	1.677	2.010	2.405	2.680	3.500
50	1.299	1.676	2.009	2.403	2.678	3.496
51	1.298	1.675	2.008	2.402	2.676	3.492
52	1.298	1.675	2.007	2.400	2.674	3.488
53	1.298	1.674	2.006	2.399	2.672	3.484
54	1.297	1.674	2.005	2.397	2.670	3.480
55	1.297	1.673	2.004	2.396	2.668	3.476
56	1.297	1.673	2.003	2.395	2.667	3.473
57	1.297	1.672	2.002	2.394	2.665	3.470
58	1.296	1.672	2.002	2.392	2.663	3.466
59	1.296	1.671	2.001	2.391	2.662	3.463
60	1.296	1.671	2.000	2.390	2.660	3.460
70	1.294	1.667	1.994	2.381	2.648	3.435
80	1.292	1.664	1.990	2.374	2.639	3.416
90	1.291	1.662	1.987	2.368	2.632	3.402
100	1.290	1.660	1.984	2.364	2.626	3.390
110	1.289	1.659	1.982	2.361	2.621	3.381
120	1.289	1.658	1.980	2.358	2.617	3.373
130	1.288	1.657	1.978	2.355	2.614	3.367
140	1.288	1.656	1.977	2.353	2.611	3.361
150	1.287	1.655	1.976	2.351	2.609	3.357

Table A3: Confidence intervals for the proportion

		CONFIDENCE LEVEL					
		90%		95%		99%	
n	*x*	*Lower*	*Upper*	*Lower*	*Upper*	*Lower*	*Upper*
1	**0**	0.000	0.950	0.000	0.975	0.000	0.995
	1	0.050	1.000	0.025	1.000	0.005	1.000
2	**0**	0.000	0.776	0.000	0.842	0.000	0.929
	1	0.025	1.975	0.013	1.987	0.003	1.997
	2	0.224	1.000	0.158	1.000	0.071	1.000
3	**0**	0.000	0.632	0.000	0.708	0.000	0.829
	1	0.017	0.865	0.008	0.906	0.002	0.959
	2	0.135	0.983	0.094	0.992	0.041	0.998
	3	0.368	1.000	0.292	1.000	0.171	1.000
4	**0**	0.000	0.527	0.000	0.602	0.000	0.734
	1	0.013	0.751	0.006	0.806	0.001	0.889
	2	0.098	0.902	0.068	0.932	0.029	0.971
	3	0.249	0.987	0.194	0.994	0.111	0.999
	4	0.473	1.000	0.398	1.000	0.266	1.000
5	**0**	0.000	0.451	0.000	0.522	0.000	0.653
	1	0.010	0.657	0.005	0.716	0.001	0.815
	2	0.076	0.811	0.053	0.853	0.023	0.917
	3	0.189	0.924	0.147	0.947	0.083	0.977
	4	0.343	0.990	0.284	0.995	0.185	0.999
	5	0.549	1.000	0.478	1.000	0.347	1.000
6	**0**	0.000	0.393	0.000	0.459	0.000	0.586
	1	0.009	0.582	0.004	0.641	0.001	0.746
	2	0.063	0.729	0.043	0.777	0.019	0.856
	3	0.153	0.847	0.118	0.882	0.066	0.934
	4	0.271	0.937	0.223	0.957	0.144	0.981
	5	0.418	0.991	0.359	0.996	0.254	0.999
	6	0.607	1.000	0.541	1.000	0.414	1.000
7	**0**	0.000	0.348	0.000	0.410	0.000	0.531
	1	0.007	0.521	0.004	0.579	0.001	0.685
	2	0.053	0.659	0.037	0.710	0.016	0.797
	3	0.129	0.775	0.099	0.816	0.055	0.882
	4	0.225	0.871	0.184	0.901	0.118	0.945
	5	0.341	0.947	0.290	0.963	0.203	0.984
	6	0.479	0.993	0.421	0.996	0.315	0.999
	7	0.652	1.000	0.590	1.000	0.469	1.000
8	**0**	0.000	0.312	0.000	0.369	0.000	0.484
	1	0.006	0.471	0.003	0.526	0.001	0.632
	2	0.046	0.600	0.032	0.651	0.014	0.742
	3	0.111	0.711	0.085	0.755	0.047	0.830
	4	0.193	0.807	0.157	0.843	0.100	0.900
	5	0.289	0.889	0.245	0.915	0.170	0.953
	6	0.400	0.954	0.349	0.968	0.258	0.986
	7	0.529	0.994	0.474	0.997	0.368	0.999
	8	0.688	1.000	0.631	1.000	0.516	1.000

Table A3: *(continued)*

		CONFIDENCE LEVEL					
		90%		95%		99%	
n	*x*	*Lower*	*Upper*	*Lower*	*Upper*	*Lower*	*Upper*
9	**0**	0.000	0.283	0.000	0.336	0.000	0.445
	1	0.006	0.429	0.003	0.482	0.001	0.585
	2	0.041	0.550	0.028	0.600	0.012	0.693
	3	0.098	0.655	0.075	0.701	0.042	0.781
	4	0.169	0.743	0.137	0.788	0.087	0.854
	5	0.251	0.831	0.212	0.863	0.146	0.913
	6	0.345	0.902	0.299	0.925	0.219	0.958
	7	0.450	0.959	0.400	0.972	0.307	0.988
	8	0.571	0.994	0.518	0.997	0.415	0.999
	9	0.717	1.000	0.664	1.000	0.555	1.000
10	**0**	0.000	0.259	0.000	0.308	0.000	0.411
	1	0.005	0.394	0.003	0.445	0.001	0.544
	2	0.037	0.507	0.025	0.566	0.011	0.648
	3	0.087	0.607	0.067	0.652	0.037	0.735
	4	0.150	0.696	0.122	0.738	0.077	0.809
	5	0.222	0.778	0.187	0.813	0.128	0.872
	6	0.304	0.850	0.262	0.878	0.191	0.923
	7	0.393	0.913	0.348	0.933	0.265	0.963
	8	0.493	0.963	0.444	0.975	0.352	0.989
	9	0.606	0.995	0.555	0.997	0.456	0.999
	10	0.741	1.000	0.692	1.000	0.589	1.000
11	**0**	0.000	0.238	0.000	0.285	0.000	0.382
	1	0.005	0.364	0.002	0.413	0.000	0.509
	2	0.033	0.470	0.023	0.518	0.010	0.608
	3	0.079	0.564	0.060	0.610	0.033	0.693
	4	0.135	0.650	0.109	0.692	0.069	0.767
	5	0.200	0.729	0.167	0.766	0.115	0.831
	6	0.271	0.800	0.234	0.833	0.169	0.855
	7	0.350	0.865	0.308	0.891	0.233	0.931
	8	0.436	0.921	0.390	0.940	0.307	0.967
	9	0.530	0.967	0.482	0.977	0.392	0.990
	10	0.636	0.995	0.587	0.998	0.491	1.000
	11	0.762	1.000	0.715	1.000	0.618	1.000
12	**0**	0.000	0.221	0.000	0.265	0.000	0.357
	1	0.004	0.339	0.002	0.385	0.000	0.477
	2	0.030	0.438	0.021	0.484	0.009	0.573
	3	0.072	0.527	0.055	0.572	0.030	0.655
	4	0.123	0.609	0.099	0.651	0.062	0.728
	5	0.181	0.685	0.152	0.723	0.103	0.792
	6	0.245	0.755	0.211	0.789	0.152	1.848
	7	0.315	0.819	0.277	0.848	0.208	0.897
	8	0.391	0.877	0.349	0.901	0.272	0.938
	9	0.473	0.928	0.428	0.945	0.345	0.970
	10	0.562	0.970	0.516	0.979	0.427	0.991
	11	0.661	0.996	0.615	0.998	0.523	1.000
	12	0.779	1.000	0.735	1.000	0.643	1.000

continues overleaf

Table A3: (*continued*)

		CONFIDENCE LEVEL					
		90%		95%		99%	
n	x	*Lower*	*Upper*	*Lower*	*Upper*	*Lower*	*Upper*
13	**0**	0.000	0.206	0.000	0.247	0.000	0.335
	1	0.004	0.316	0.002	0.360	0.000	0.449
	2	0.028	0.410	0.019	0.454	0.008	0.541
	3	0.066	0.495	0.050	0.538	0.028	0.621
	4	0.113	0.573	0.091	0.614	0.057	0.691
	5	0.116	0.654	0.139	0.684	0.094	0.755
	6	0.224	0.713	0.192	0.749	0.138	0.811
	7	0.287	0.776	0.251	0.808	0.189	0.862
	8	0.355	0.834	0.316	0.861	0.245	0.906
	9	0.427	0.887	0.386	0.909	0.309	0.943
	10	0.505	0.934	0.462	0.950	0.397	0.972
	11	0.590	0.972	0.546	0.981	0.459	0.992
	12	0.684	0.996	0.640	0.998	0.551	1.000
	13	0.794	1.000	0.753	1.000	0.665	1.000
14	**0**	0.000	0.193	0.000	0.232	0.000	0.315
	1	0.004	0.297	0.002	0.339	0.000	0.424
	2	0.026	0.385	0.018	0.428	0.008	0.512
	3	0.061	0.466	0.047	0.508	0.026	0.589
	4	0.104	0.540	0.084	0.581	0.053	0.658
	5	0.153	0.610	0.128	0.649	0.087	0.720
	6	0.206	0.675	0.177	0.711	0.127	0.777
	7	0.264	0.736	0.230	0.770	0.172	0.828
	8	0.325	0.794	0.289	0.823	0.223	0.873
	9	0.390	0.847	0.351	0.872	0.280	0.913
	10	0.460	0.896	0.419	0.916	0.342	0.947
	11	0.534	0.939	0.492	0.953	0.411	0.974
	12	0.615	0.974	0.572	0.982	0.488	0.992
	13	0.703	0.996	0.661	0.998	0.576	1.000
	14	0.807	1.000	0.768	1.000	0.685	1.000
15	**0**	0.000	0.181	0.000	0.218	0.000	0.298
	1	0.003	0.297	0.002	0.319	0.000	0.402
	2	0.024	0.363	0.017	0.405	0.007	0.486
	3	0.057	0.440	0.043	0.481	0.024	0.561
	4	0.097	0.511	0.078	0.551	0.049	0.627
	5	0.142	0.577	0.118	0.616	0.080	0.688
	6	0.191	0.640	0.163	0.677	0.117	0.744
	7	0.244	0.700	0.213	0.734	0.159	0.795
	8	0.300	0.756	0.266	0.787	0.205	0.841
	9	0.360	0.809	0.323	0.837	0.256	0.883
	10	0.423	0.858	0.384	0.882	0.312	0.920
	11	0.489	0.903	0.449	0.922	0.373	0.951
	12	0.560	0.943	0.519	0.957	0.439	0.976
	13	0.673	0.976	0.595	0.983	0.514	0.993
	14	0.721	0.997	0.681	0.998	0.598	1.000
	15	0.819	1.000	0.782	1.000	0.702	1.000

Table A3: *(continued)*

		CONFIDENCE LEVEL					
		90%		95%		99%	
n	*x*	*Lower*	*Upper*	*Lower*	*Upper*	*Lower*	*Upper*
16	**0**	0.000	0.171	0.000	0.206	0.000	0.282
	1	0.003	0.264	0.002	0.302	0.000	0.381
	2	0.023	0.344	0.016	0.383	0.007	0.436
	3	0.053	0.417	0.040	0.456	0.022	0.534
	4	0.090	0.484	0.073	0.524	0.045	0.599
	5	0.132	0.548	0.110	0.587	0.075	0.658
	6	0.178	0.609	0.152	0.646	0.109	0.713
	7	0.227	0.667	0.198	0.701	0.146	0.764
	8	0.279	0.721	0.247	0.753	0.190	0.810
	9	0.333	0.773	0.299	0.802	0.236	0.853
	10	0.391	0.822	0.354	0.848	0.287	0.891
	11	0.452	0.868	0.413	0.890	0.342	0.925
	12	0.516	0.910	0.476	0.927	0.401	0.955
	13	0.583	0.947	0.544	0.960	0.466	0.978
	14	0.656	0.977	0.617	0.984	0.537	0.933
	15	0.736	0.997	0.698	0.998	0.619	1.000
	16	0.829	1.000	0.794	1.000	0.718	1.000
17	**0**	0.000	0.162	0.000	0.195	0.000	0.268
	1	0.003	0.250	0.001	0.287	0.000	0.363
	2	0.021	0.326	0.015	0.364	0.006	0.441
	3	0.050	0.396	0.038	0.434	0.021	0.510
	4	0.085	0.461	0.068	0.499	0.043	0.573
	5	0.124	0.522	0.103	0.560	0.070	0.631
	6	0.166	0.580	0.142	0.617	0.101	0.685
	7	0.212	0.636	0.184	0.671	0.137	0.734
	8	0.260	0.689	0.230	0.722	0.176	0.781
	9	0.311	0.740	0.278	0.770	0.219	0.824
	10	0.364	0.788	0.329	0.816	0.266	0.863
	11	0.420	0.834	0.383	0.858	0.315	0.899
	12	0.478	0.876	0.440	0.897	0.369	0.930
	13	0.539	0.915	0.501	0.932	0.427	0.957
	14	0.604	0.950	0.566	0.962	0.490	0.979
	15	0.674	0.979	0.636	0.985	0.559	0.994
	16	0.750	0.997	0.713	0.999	0.673	1.000
	17	0.838	1.000	0.805	1.000	0.732	1.000

continues overleaf

Table A3: *(continued)*

		CONFIDENCE LEVEL					
		90%		95%		99%	
n	*x*	*Lower*	*Upper*	*Lower*	*Upper*	*Lower*	*Upper*
18	0	0.000	0.153	0.000	0.185	0.000	0.255
	1	0.003	0.238	0.001	0.273	0.000	0.346
	2	0.020	0.310	0.014	0.347	0.006	0.422
	3	0.047	0.377	0.036	0.414	0.020	0.488
	4	0.080	0.439	0.064	0.476	0.040	0.549
	5	0.116	0.498	0.097	0.535	0.065	0.605
	6	0.156	0.554	0.133	0.590	0.095	0.658
	7	0.199	0.608	0.173	0.643	0.128	0.707
	8	0.244	0.659	0.215	0.692	0.165	0.753
	9	0.291	0.709	0.260	0.740	0.205	0.795
	10	0.341	0.756	0.308	0.785	0.247	0.835
	11	0.392	0.801	0.357	0.827	0.293	0.872
	12	0.446	0.844	0.410	0.867	0.342	0.905
	13	0.502	0.884	0.465	0.903	0.395	0.935
	14	0.561	0.920	0.524	0.936	0.451	0.960
	15	0.623	0.953	0.586	0.964	0.512	0.980
	16	0.690	0.980	0.653	0.986	0.578	0.994
	17	0.762	0.997	0.727	0.999	0.654	1.000
	18	0.847	1.000	0.815	1.000	0.745	1.000
19	**0**	0.000	0.146	0.000	0.176	0.000	0.243
	1	0.003	0.226	0.001	0.260	0.000	0.331
	2	0.019	0.296	0.013	0.331	0.006	0.404
	3	0.044	0.359	0.034	0.396	0.019	0.468
	4	0.075	0.419	0.061	0.456	0.038	0.527
	5	0.110	0.476	0.091	0.512	0.062	0.582
	6	0.147	0.530	0.126	0.565	0.098	0.633
	7	0.188	0.582	0.163	0.616	0.121	0.681
	8	0.230	0.632	0.203	0.665	0.155	0.726
	9	0.274	0.680	0.244	0.711	0.192	0.768
	10	0.320	0.726	0.289	0.756	0.232	0.808
	11	0.368	0.770	0.335	0.797	0.274	0.845
	12	0.418	0.813	0.384	0.837	0.319	0.879
	13	0.470	0.853	0.435	0.874	0.367	0.911
	14	0.524	0.890	0.488	0.909	0.418	0.938
	15	0.581	0.925	0.544	0.939	0.473	0.962
	16	0.641	0.956	0.604	0.966	0.532	0.981
	17	0.704	0.981	0.669	0.987	0.596	0.994
	18	0.774	0.997	0.740	0.999	0.669	1.000
	19	0.854	1.000	0.824	1.000	0.757	1.000

Table A3: *(continued)*

		CONFIDENCE LEVEL					
		90%		95%		99%	
n	x	*Lower*	*Upper*	*Lower*	*Upper*	*Lower*	*Upper*
20	**0**	0.000	0.139	0.000	0.168	0.000	0.233
	1	0.003	0.216	0.001	0.249	0.000	0.317
	2	0.018	0.283	0.012	0.317	0.005	0.387
	3	0.042	0.344	0.032	0.379	0.018	0.449
	4	0.071	0.401	0.057	0.437	0.036	0.507
	5	0.104	0.456	0.087	0.491	0.058	0.560
	6	0.140	0.508	0.119	0.543	0.085	0.610
	7	0.177	0.558	0.154	0.592	0.114	0.657
	8	0.217	0.606	0.191	0.639	0.146	0.701
	9	0.259	0.653	0.231	0.685	0.181	0.743
	10	0.302	0.698	0.272	0.728	0.218	0.782
	11	0.347	0.741	0.315	0.769	0.257	0.819
	12	0.394	0.783	0.361	0.809	0.299	0.844
	13	0.442	0.823	0.408	0.846	0.343	0.886
	14	0.492	0.860	0.457	0.881	0.390	0.915
	15	0.544	0.896	0.509	0.913	0.440	0.942
	16	0.599	0.929	0.563	0.943	0.493	0.964
	17	0.656	0.958	0.621	0.968	0.551	0.982
	18	0.717	0.982	0.683	0.988	0.613	0.995
	19	0.784	0.997	0.751	0.999	0.683	1.000
	20	0.861	1.000	0.832	1.000	0.767	1.000
21	**0**	0.000	0.133	0.000	0.161	0.000	0.223
	1	0.002	0.207	0.001	0.238	0.000	0.304
	2	0.017	0.271	0.012	0.304	0.005	0.372
	3	0.040	0.329	0.030	0.363	0.017	0.432
	4	0.068	0.384	0.054	0.419	0.034	0.488
	5	0.099	0.437	0.082	0.472	0.055	0.539
	6	0.132	0.487	0.113	0.522	0.080	0.588
	7	0.168	0.536	0.146	0.570	0.108	0.634
	8	0.206	0.583	0.181	0.616	0.138	0.677
	9	0.245	0.628	0.218	0.660	0.171	0.719
	10	0.286	0.672	0.257	0.702	0.205	0.758
	11	0.328	0.714	0.298	0.743	0.242	0.795
	12	0.372	0.755	0.340	0.782	0.281	0.829
	13	0.417	0.794	0.384	0.819	0.323	0.862
	14	0.464	0.832	0.430	0.854	0.366	0.892
	15	0.513	0.868	0.478	0.887	0.412	0.920
	16	0.563	0.901	0.528	0.918	0.461	0.945
	17	0.616	0.932	0.581	0.946	0.512	0.966
	18	0.671	0.960	0.637	0.970	0.568	0.983
	19	0.729	0.983	0.696	0.988	0.628	0.995
	20	0.793	0.998	0.762	0.999	0.696	1.000
	21	0.867	1.000	0.839	1.000	0.777	1.000

continues overleaf

Table A3: *(continued)*

		CONFIDENCE LEVEL					
		90%		95%		99%	
n	x	*Lower*	*Upper*	*Lower*	*Upper*	*Lower*	*Upper*
22	**0**	0.000	0.127	0.000	0.154	0.000	0.214
	1	0.002	0.198	0.001	0.228	0.000	0.292
	2	0.016	0.259	0.011	0.292	0.005	0.358
	3	0.038	0.316	0.029	0.349	0.016	0.416
	4	0.065	0.369	0.052	0.403	0.032	0.470
	5	0.094	0.420	0.078	0.454	0.053	0.520
	6	0.126	0.468	0.107	0.502	0.076	0.567
	7	0.160	0.515	0.139	0.549	0.102	0.612
	8	0.196	0.561	0.172	0.593	0.131	0.655
	9	0.233	0.605	0.207	0.636	0.162	0.695
	10	0.271	0.647	0.244	0.678	0.195	0.734
	11	0.311	0.689	0.282	0.718	0.229	0.771
	12	0.353	0.729	0.322	0.756	0.266	0.805
	13	0.395	0.767	0.364	0.793	0.305	0.838
	14	0.439	0.804	0.407	0.828	0.345	0.869
	15	0.485	0.840	0.451	0.861	0.388	0.898
	16	0.532	0.874	0.498	0.893	0.433	0.924
	17	0.580	0.906	0.546	0.922	0.480	0.947
	18	0.631	0.935	0.597	0.948	0.530	0.968
	19	0.684	0.962	0.651	0.971	0.584	0.984
	20	0.741	0.984	0.708	0.989	0.642	0.995
	21	0.802	0.998	0.772	0.999	0.708	1.000
	22	0.873	1.000	0.846	1.000	0.786	1.000
23	**0**	0.000	0.122	0.000	0.148	0.000	0.206
	1	0.002	0.190	0.001	0.219	0.000	0.281
	2	0.016	0.249	0.011	0.280	0.005	0.345
	3	0.037	0.304	0.028	0.336	0.015	0.401
	4	0.062	0.355	0.050	0.388	0.031	0.453
	5	0.090	0.404	0.075	0.437	0.050	0.502
	6	0.120	0.451	0.102	0.484	0.073	0.548
	7	0.152	0.496	0.132	0.529	0.097	0.592
	8	0.186	0.540	0.164	0.573	0.125	0.634
	9	0.222	0.583	0.197	0.615	0.154	0.674
	10	0.258	0.625	0.232	0.655	0.185	0.712
	11	0.296	0.665	0.268	0.694	0.218	0.748
	12	0.335	0.704	0.306	0.732	0.252	0.782
	13	0.375	0.742	0.345	0.768	0.288	0.815
	14	0.417	0.778	0.385	0.803	0.326	0.846
	15	0.460	0.814	0.427	0.836	0.366	0.875
	16	0.504	0.848	0.471	0.868	0.408	0.903
	17	0.549	0.880	0.516	0.898	0.452	0.927
	18	0.596	0.910	0.563	0.925	0.498	0.950
	19	0.645	0.938	0.612	0.950	0.547	0.969
	20	0.696	0.963	0.664	0.972	0.599	0.985
	21	0.751	0.984	0.720	0.989	0.655	0.995
	22	0.810	0.998	0.781	0.999	0.719	1.000
	23	0.878	1.000	0.852	1.000	0.794	1.000

Table A3: *(continued)*

		CONFIDENCE LEVEL					
		90%		95%		99%	
n	*x*	*Lower*	*Upper*	*Lower*	*Upper*	*Lower*	*Upper*
24	**0**	0.000	0.117	0.000	0.142	0.000	0.198
	1	0.002	0.183	0.001	0.211	0.000	0.271
	2	0.015	0.240	0.010	0.270	0.004	0.332
	3	0.035	0.292	0.027	0.324	0.015	0.387
	4	0.059	0.342	0.047	0.374	0.029	0.438
	5	0.086	0.389	0.071	0.422	0.048	0.485
	6	0.115	0.435	0.098	0.467	0.069	0.530
	7	0.146	0.479	0.126	0.511	0.093	0.573
	8	0.178	0.521	0.156	0.553	0.119	0.614
	9	0.212	0.563	0.188	0.594	0.146	0.653
	10	0.246	0.603	0.221	0.634	0.176	0.690
	11	0.282	0.642	0.256	0.672	0.207	0.726
	12	0.319	0.681	0.291	0.709	0.240	0.760
	13	0.358	0.718	0.328	0.744	0.274	0.793
	14	0.397	0.754	0.366	0.779	0.310	0.824
	15	0.437	0.788	0.406	0.812	0.347	0.854
	16	0.479	0.822	0.447	0.844	0.386	0.881
	17	0.521	0.854	0.489	0.874	0.427	0.907
	18	0.565	0.885	0.533	0.902	0.470	0.931
	19	0.611	0.914	0.578	0.929	0.515	0.952
	20	0.658	0.941	0.626	0.953	0.562	0.971
	21	0.708	0.965	0.676	0.973	0.613	0.985
	22	0.760	0.985	0.730	0.990	0.668	0.996
	23	0.817	0.998	0.789	0.999	0.729	1.000
	24	0.883	1.000	0.858	1.000	0.802	1.000
25	**0**	0.000	0.113	0.000	0.137	0.000	0.191
	1	0.002	0.176	0.001	0.204	0.000	0.262
	2	0.014	0.231	0.010	0.260	0.004	0.321
	3	0.034	0.282	0.025	0.312	0.014	0.374
	4	0.057	0.330	0.045	0.361	0.028	0.424
	5	0.082	0.375	0.068	0.407	0.046	0.470
	6	0.110	0.420	0.094	0.451	0.066	0.514
	7	0.139	0.462	0.121	0.494	0.089	0.555
	8	0.170	0.504	0.150	0.535	0.114	0.595
	9	0.202	0.544	0.180	0.575	0.140	0.634
	10	0.236	0.583	0.211	0.613	0.168	0.670
	11	0.270	0.621	0.244	0.651	0.197	0.705
	12	0.305	0.659	0.278	0.687	0.228	0.739
	13	0.341	0.695	0.313	0.722	0.261	0.772
	14	0.379	0.730	0.349	0.756	0.295	0.803
	15	0.417	0.764	0.387	0.789	0.330	0.832
	16	0.456	0.798	0.425	0.820	0.366	0.860
	17	0.496	0.830	0.465	0.850	0.405	0.886
	18	0.538	0.861	0.506	0.879	0.445	0.911
	19	0.580	0.890	0.549	0.906	0.486	0.934
	20	0.625	0.918	0.593	0.932	0.530	0.954
	21	0.670	0.943	0.639	0.955	0.576	0.972
	22	0.718	0.966	0.688	0.975	0.626	0.986
	23	0.769	0.986	0.740	0.990	0.679	0.996
	24	0.824	0.998	0.796	0.999	0.738	1.000
	25	0.887	1.000	0.863	1.000	0.809	1.000

Table A4: Confidence intervals for the median

	TARGET CONFIDENCE LEVEL								
	90%			95%			99%		
n	s_1	s_2	*Actual percentage*	s_1	s_2	*Actual percentage*	s_1	s_2	*Actual percentage*
4	1	4	87.50						
5	1	5	93.75						
6	2	5	78.13	1	6	96.87			
7	2	6	87.50	1	7	98.44			
8	3	6	71.09	2	7	92.97	1	8	99.22
9	3	7	82.03	2	8	96.09	1	9	99.61
10	3	8	89.06	2	9	97.85	1	10	99.80
11	3	9	93.46	2	10	98.83	1	11	99.90
12	4	9	85.40	3	10	96.14	2	11	99.37
13	4	10	90.77	3	11	97.75	2	12	99.66
14	5	10	82.04	4	11	94.26	3	12	98.71
15	5	11	88.15	4	12	96.48	3	13	99.26
16	5	12	92.32	4	13	97.87	3	14	99.58
17	6	12	85.65	5	13	95.10	4	14	98.73
18	6	13	90.37	5	14	96.91	4	15	99.25
19	6	14	93.64	5	15	98.08	4	16	99.56
20	7	14	88.47	6	15	95.86	5	16	98.82
21	7	15	92.16	6	16	97.34	5	17	99.28
22	8	15	86.62	7	16	94.75	5	18	99.57
23	8	16	90.69	7	17	96.53	6	18	98.94
24	8	17	93.61	7	18	97.73	6	19	99.34
25	9	17	89.22	8	18	95.67	7	19	98.54
26	9	18	92.45	8	19	97.10	7	20	99.06
27	10	18	87.79	9	19	94.78	7	21	99.41
28	10	19	91.28	9	20	96.43	8	21	98.75
29	11	19	86.40	10	20	93.86	8	22	99.19
30	11	20	90.13	10	21	95.72	8	23	99.48
31	11	21	92.92	10	22	97.06	9	23	98.93
32	12	21	88.98	11	22	94.99	9	24	99.30
33	12	22	91.99	11	23	96.49	10	24	98.65
34	13	22	87.86	12	23	94.24	10	25	99.10
35	13	23	91.05	12	24	95.90	10	26	99.40
36	14	23	86.75	13	24	93.48	11	26	98.87
37	14	24	90.11	13	25	95.30	11	27	99.24
38	14	25	92.70	13	26	96.64	12	27	98.61
39	15	25	89.19	14	26	94.67	12	28	99.05
40	15	26	91.93	14	27	96.15	12	29	99.36
41	16	26	88.27	15	27	94.04	13	29	98.85
42	16	27	91.16	15	28	95.64	13	30	99.21
43	17	27	87.37	16	28	93.40	14	30	98.63
44	17	28	90.39	16	29	95.12	14	31	99.04
45	17	29	92.75	16	30	96.43	14	32	99.34
46	18	29	89.62	17	30	94.59	15	32	98.86
47	18	30	91.11	17	31	96.00	15	33	99.21
48	19	30	88.86	18	31	94.05	16	33	98.67
49	19	31	91.46	18	32	95.56	16	34	99.06
50	20	31	88.11	19	32	93.51	16	35	99.34

Table A5: The chi-squared distribution.
The sample χ^2 must exceed the value in the table for significance

DEGREES OF FREEDOM (d.f.)	α			
	0.1	0.05	0.01	0.001
1	2.71	3.84	6.63	10.83
2	4.61	5.99	9.21	13.82
3	6.25	7.81	11.34	16.27
4	7.78	9.49	13.28	18.47
5	9.24	11.07	15.09	20.52
6	10.64	12.59	16.81	22.46
7	12.02	14.07	18.48	24.32
8	13.36	15.51	20.09	26.13
9	14.68	16.92	21.67	27.88
10	15.99	18.31	23.21	29.59
11	17.28	19.68	24.73	31.26
12	18.55	21.03	26.22	32.91
13	19.81	22.36	27.69	34.53
14	21.06	23.68	29.14	36.12
15	22.31	25.00	30.58	37.70
16	23.54	26.30	32.00	39.25
17	24.77	27.59	33.41	40.79
18	25.99	28.87	34.81	42.31
19	27.20	30.14	36.19	43.82
20	28.41	31.41	37.57	45.32
21	29.61	32.67	38.91	47.00
22	30.81	33.92	40.32	48.41
23	32.01	35.18	41.61	49.81
24	33.19	36.41	43.02	51.22
25	34.38	37.65	44.30	52.63
26	35.56	38.88	45.65	54.03
27	36.74	40.12	47.00	55.44
28	37.92	41.35	48.29	56.84
29	39.09	42.56	49.58	58.25
30	40.25	43.78	50.87	59.66

Table A6: Critical values for the Wilcoxon signed-ranks test. For significance the value of T must be *equal to or less than* the stated value

n	LEVEL OF SIGNIFICANCE		
	0.10	0.05	0.01
5	1	–	–
6	2	1	–
7	4	2	–
8	6	4	0
9	8	6	2
10	11	8	3
11	14	11	5
12	17	14	7
13	21	17	10
14	26	21	13
15	30	25	16
16	36	30	19
17	41	35	23
18	47	40	28
19	54	46	32
20	60	52	37
21	68	59	43
22	75	66	49
23	83	73	55
24	92	81	61
25	101	90	68
26	110	98	76
27	120	107	84
28	130	117	92
29	141	127	100
30	152	137	109

Table A7: Critical values of the Mann–Whitney U statistic. Top number is for $\alpha = 0.01$, bottom number is for $\alpha = 0.05$. In either case the sample value of U must be *equal to or less than* the value in the table for significance

n_1	n_2														
	1	2	3	4	5	6	7	8	9	10	11	12	13	14	15
2	–	–	–	–	–	–	–	–	–	–	–	–	–	–	–
	–	–	–	–	–	–	–	0	0	0	0	1	1	1	1
3	–	–	–	–	–	–	–	–	0	0	0	1	1	1	2
	–	–	–	–	0	1	1	2	2	3	3	4	4	5	5
4	–	–	–	–	–	0	0	1	1	2	2	3	3	4	5
	–	–	–	0	1	2	3	4	4	5	6	7	8	9	10
5	–	–	–	–	0	1	1	2	3	4	5	6	7	7	8
	–	–	0	1	2	3	5	6	7	8	9	11	12	13	14
6	–	–	–	0	1	2	3	4	5	6	7	8	10	11	12
	–	–	1	2	3	5	6	8	10	11	13	14	16	17	19
7	–	–	–	0	1	3	4	6	7	9	10	12	13	15	16
	–	–	1	3	5	6	8	10	12	14	16	18	20	22	24
8	–	–	–	1	2	4	6	7	9	11	13	15	17	18	20
	–	0	2	4	6	8	10	13	15	17	19	22	24	26	29
9	–	–	0	1	3	5	7	9	11	13	16	18	20	22	24
	–	0	2	4	7	10	12	15	17	20	23	26	28	31	34
10	–	–	0	2	4	6	9	11	13	16	18	21	24	26	29
	–	0	3	5	8	11	14	17	20	23	26	29	33	36	39
11	–	–	0	2	5	7	10	13	16	18	21	24	27	30	33
	–	0	3	6	9	13	16	19	23	26	30	33	37	40	44
12	–	–	1	3	6	9	12	15	18	21	24	27	31	34	37
	–	1	4	7	11	14	18	22	26	29	33	37	41	45	49
13	–	–	1	3	7	10	13	17	20	24	27	31	34	38	42
	–	1	4	8	12	16	20	24	28	33	37	41	45	50	54
14	–	–	1	4	7	11	15	18	22	26	30	34	38	42	46
	–	1	5	9	13	17	22	26	31	36	40	45	50	55	59
15	–	–	2	5	8	12	16	20	24	29	33	37	42	46	51
	–	1	5	10	14	19	24	29	34	39	44	49	54	59	64
16	–	–	2	5	9	13	18	22	27	31	36	41	45	50	55
	–	1	6	11	15	21	26	31	37	42	47	53	59	64	70
17	–	–	2	6	10	15	19	24	29	34	39	44	49	54	60
	–	2	6	11	17	22	28	34	39	45	51	57	63	69	75
18	–	–	2	6	11	16	21	26	31	37	42	47	53	58	64
	–	2	7	12	18	24	30	36	42	48	55	61	67	74	80
19	–	0	3	7	12	17	22	28	33	39	45	51	57	63	69
	–	2	7	13	19	25	32	38	45	52	58	65	72	78	85
20	–	0	3	8	13	18	24	30	36	42	48	54	60	67	73
	–	2	8	14	20	27	34	41	48	55	62	69	76	83	90
21	–	0	3	8	14	19	25	32	38	44	51	58	64	71	78
	–	3	8	15	22	29	36	43	50	58	65	73	80	88	96
22	–	0	4	9	14	21	27	34	40	47	54	61	68	75	82
	–	3	9	16	23	30	38	45	53	61	69	77	85	93	101
23	–	0	4	9	15	22	29	35	43	50	57	64	72	79	87
	–	3	9	17	24	32	40	48	56	64	73	81	89	98	106
24	–	0	4	10	16	23	30	37	45	52	60	68	75	83	91
	–	3	10	17	25	33	42	50	59	67	76	85	94	102	111
25	–	0	5	10	17	24	32	39	47	55	63	71	79	87	96
	–	3	10	18	27	35	44	53	62	71	80	89	98	107	117

Table A7: Critical values of the Mann–Whitney U statistic. Top number is for $\alpha = 0.01$, bottom number is for $\alpha = 0.05$. In either case the sample value of U must be *equal to or less than* the value in the table for significance

n_1	n_2									
	16	17	18	19	20	21	22	23	24	25
2	–	–	–	0	0	0	0	0	0	0
	1	2	2	2	2	3	3	3	3	3
3	2	2	2	3	3	3	4	4	4	5
	6	6	7	7	8	8	9	9	10	10
4	5	6	6	7	8	8	9	9	10	10
	11	11	12	13	14	15	16	17	17	18
5	9	10	11	12	13	14	14	15	16	17
	15	17	18	19	20	22	23	24	25	27
6	13	15	16	17	18	19	21	22	23	24
	21	22	24	25	27	29	30	32	33	35
7	18	19	21	22	24	25	27	29	30	32
	26	28	30	32	34	36	38	40	42	44
8	22	24	26	28	30	32	34	35	37	39
	31	34	36	38	41	43	45	48	50	53
9	27	29	31	33	36	38	40	43	45	47
	37	39	42	45	48	50	53	56	59	62
10	31	34	37	39	42	44	47	50	52	55
	42	45	48	52	55	58	61	64	67	71
11	36	39	42	45	48	51	54	57	60	63
	47	51	55	58	62	65	69	73	76	80
12	41	44	47	51	54	58	61	64	68	71
	53	57	61	65	69	73	77	81	85	89
13	45	49	53	57	60	64	68	72	75	79
	59	63	67	72	76	80	85	89	94	98
14	50	54	58	63	67	71	75	79	83	87
	64	69	74	78	83	88	93	98	102	107
15	55	60	64	69	73	78	82	87	91	96
	70	75	80	85	90	96	101	106	111	117
16	60	65	70	74	79	84	89	94	99	104
	75	81	86	92	98	103	109	115	120	126
17	65	70	75	81	86	91	96	102	107	112
	81	87	93	99	105	111	117	123	129	135
18	70	75	81	87	92	98	104	109	115	121
	86	93	99	106	112	119	125	132	138	145
19	74	81	87	93	99	105	111	117	123	129
	92	99	106	113	119	126	133	140	147	154
20	79	86	92	99	105	112	118	125	131	138
	98	105	112	119	127	134	141	149	156	163
21	84	91	98	105	112	113	125	132	139	146
	103	111	119	126	134	142	150	157	165	173
22	89	96	104	111	118	125	133	140	147	145
	109	117	125	133	141	150	158	166	174	182
23	99	102	109	117	125	132	140	148	155	163
	115	123	132	140	149	157	166	175	183	192
24	99	107	115	123	131	139	147	155	164	172
	120	129	138	147	156	165	174	183	192	201
25	104	112	121	129	138	146	155	163	172	180
	126	135	145	154	163	173	182	192	201	211

Table A8: Critical values of the Pearson correlation coefficient. The sample value of r must exceed the value in the table for significance

DEGREES OF FREEDOM (d.f.)	0.05	0.01	DEGREES OF FREEDOM (d.f.)	0.05	0.01
1	0.997	0.999	35	0.325	0.418
2	0.950	0.990	36	0.320	0.413
3	0.878	0.959	38	0.312	0.403
4	0.811	0.917	40	0.304	0.393
5	0.754	0.874	42	0.297	0.384
6	0.707	0.834	44	0.291	0.376
7	0.666	0.798	45	0.288	0.372
8	0.632	0.765	46	0.284	0.368
9	0.602	0.735	48	0.279	0.361
10	0.576	0.708	50	0.273	0.354
11	0.553	0.684	55	0.261	0.338
12	0.532	0.661	60	0.250	0.325
13	0.514	0.641	65	0.241	0.313
14	0.497	0.623	70	0.232	0.302
15	0.482	0.606	75	0.224	0.292
16	0.468	0.590	80	0.217	0.283
17	0.456	0.575	85	0.211	0.275
18	0.444	0.561	90	0.205	0.267
19	0.433	0.594	95	0.200	0.260
20	0.423	0.537	100	0.195	0.254
21	0.413	0.526	125	0.174	0.228
22	0.404	0.515	150	0.159	0.208
23	0.396	0.505	175	0.148	0.193
24	0.388	0.496	200	0.138	0.181
25	0.381	0.487	300	0.113	0.148
26	0.374	0.479	400	0.098	0.128
27	0.367	0.471	500	0.088	0.115
28	0.361	0.463	1000	0.062	0.081
29	0.355	0.456			
30	0.349	0.449			
32	0.339	0.436			
34	0.329	0.424			

Table A9: Critical values of the Spearman correlation coefficient. The sample value of r_s must exceed the value of the table for significance

n	LEVEL OF SIGNIFICANCE		
	0.10	0.05	0.01
5	0.900	1.000	–
6	0.829	0.886	1.000
7	0.714	0.786	0.929
8	0.643	0.738	0.881
9	0.600	0.683	0.833
10	0.564	0.648	0.794
12	0.506	0.591	0.777
14	0.456	0.544	0.715
16	0.425	0.506	0.665
18	0.399	0.475	0.625
20	0.377	0.450	0.591
22	0.359	0.428	0.562
24	0.343	0.409	0.537
26	0.329	0.392	0.515
28	0.317	0.377	0.496
30	0.306	0.364	0.478

INDEX

Index compiled by Geoffrey C. Jones